Alfred Niaudet

Elementary Treatise on Electric Batteries

Third Edition

Alfred Niaudet

Elementary Treatise on Electric Batteries
Third Edition

ISBN/EAN: 9783337275723

Printed in Europe, USA, Canada, Australia, Japan

Cover: Foto ©berggeist007 / pixelio.de

More available books at **www.hansebooks.com**

ELEMENTARY TREATISE

ON

ELECTRIC BATTERIES.

FROM THE FRENCH OF
ALFRED NIAUDET,

TRANSLATED BY
L. M. FISHBACK,
OF THE BELL TELEPHONE CO. OF MISSOURI.

THIRD EDITION.

NEW YORK:
JOHN WILEY & SONS,
15 ASTOR PLACE.
1884.

PREFACE.

The English translation of Mr. Alfred Niaudet's "La Pile Electrique" scarcely requires my commendation to render it acceptable to the English-speaking community interested in the subject, since the author's name is so well known to electricians.

This work will serve to guide the uninitiated in the choice and management of batteries, and even the professional electrician may find not only new matter but even old material presented in a new form, and worked to new developments.

Telegraphers generally will find many of their frequently recurring problems solved in its pages, and its perspicuity will save both inventor and investigator from making useless experiments or errors, while at the same time the work offers to all new fields for careful research.

Although the subject treated is so useful and interesting, yet this is, I believe, the first time it has received such recognition in English as its importance demands.

The translator was happily fitted for his task, having studied under the direction of the author himself, and with whose sanction he undertook his task.

Geo. d'Infreville,
Electrician, Western Union Telegraph Co.

New York, July 23, 1880.

PREFACE TO THE ENGLISH EDITION.

The work which we here present to the public is in conformity with the second French edition of a book the first edition of which appeared in 1878, and which has been exhausted in less than two years.

No other treatise upon the "Electric Battery" has hitherto been published either in English, French or German. It has appeared desirable to meet this need, and to offer a complete guide to those who wish to thoroughly study or even to improve upon batteries, which are to-day so extensively applied to different uses.

The order that the author has adopted in his exposition is in some sense obligatory. Single-liquid batteries are the first, historically and logically, to present themselves. In connection with this first part are naturally placed the exposition of principles, definitions of terms, and the study of the phenomenon of polarization, wherein lies the whole difficulty of the subject.

Next in order come two-liquid batteries, in which polarization is suppressed or reduced according to circumstances.

CONTENTS.

PART FIRST.
Single-Liquid Batteries.

CHAPTER I.
INTRODUCTION.

	PAGE
Definitions,	1
Origin of the Name Pile,	1
First Idea of the Battery,	2
Properties of Amalgamated Zinc,	7
Inconstancy of Simple Batteries,	8
Battery Cells joined in Intensity,	10

CHAPTER II.
DESCRIPTION OF VOLTA'S BATTERY AND ITS DERIVATIVES.

Column Battery,	13
Volta's "*Couronne de Tasses*,"	13
Cruikshank's Battery,	14
Wollaston's Battery,	15
Spiral Battery,	17
Muncke's Battery,	18
Sand Battery,	19
Nature of the Chemical Action in Volta's Battery,	20
Action of Air upon Batteries,	22

CHAPTER III.

GENERAL REMARKS UPON BATTERIES.

	PAGE
Ideas upon Electric Resistance,	23
General Remarks upon Electro-motive Force and Resistance,	24
Electro-motive Force,	26
Measurement of Electro-motive Forces,	31
Internal Resistance of the Battery,	32
Various ways of Joining Voltaic Cells,	34
The Voltameter,	38
Secondary Currents, Polarized Electrodes,	41
Polarization of a Voltaic Cell,	42
Polarization in a Battery of several Cells,	47

CHAPTER IV.

SULPHURIC-ACID BATTERIES.

Batteries with Carbon Electrodes,	49
Manufacture of Carbon Electrodes,	50
Use of Carbon Electrodes,	51
Zinc-Iron Battery,	53
Iron-Copper Battery,	53
Other Combinations,	53
Smee's Cell,	54
Walker's Platinized Carbon Battery,	56
Tyer's Battery,	57
Baron Ebner's Battery,	58
Batteries analogous to that of Smee,	59
Remarks upon Polarization in the preceding Batteries,	59

CHAPTER V.

ACID BATTERIES ANALOGOUS TO THAT OF VOLTA.

Hydrochloric-Acid Batteries,	61
Nitric-Acid Batteries,	61
Various Acid Batteries,	62

CHAPTER VI.

BATTERIES WITHOUT ACIDS.

	PAGE
Sea-salt Batteries,	63
Duchemin's Electric Buoy,	64
Sea-water, Zinc and Copper Battery,	66
Zinc, Iron, and Sea-water Battery,	67
Accidental Reversing of the Current,	68
Chemical Action in Sea-salt Batteries,	70
Marine Batteries,	71
Sal-Ammoniac Batteries,	72
Bagration Battery,	72
Carbon-Electrode Battery,	72
Action of Air upon the preceding Battery,	75
Chemical Action in Sal-Ammoniac Batteries,	76

OTHER BATTERIES.

Zinc-Iron-Water Battery,	76
Iron-Tin Battery,	77
Alum Battery,	78
Remarks upon Single-Liquid Batteries,	79

PART SECOND.

TWO-LIQUID BATTERIES.

CHAPTER I.

THE DANIELL BATTERY.

Introduction,	81
Description of the Daniell,	86
Improved Daniell Cell,	99
Balloon Battery,	101
A Reversed form of Daniell's Battery,	102
Trough Battery,	104
Conventional Figure,	106

viii   CONTENTS.

	PAGE
Muirhead's Battery,	107
Carré's Battery,	108
Siemens and Halske's Battery,	109
Varley's Battery,	110
Minotto's Battery,	111
Trouvé's Blotting-Paper Battery,	112

CHAPTER II.

GRAVITY BATTERIES.

Callaud's Battery,	118
Applications of Callaud's Battery,	122
Trouvé-Callaud Battery,	123
Meidinger's Battery,	124
Meidinger's Flask Battery,	127
Kruger's Battery,	128
Sir William Thomson's Battery,	130
Electro-motive Force of the Daniell Gravity Battery,	133

CHAPTER III.

GENERAL REMARKS UPON DANIELL BATTERIES.

Amalgamation of Zinc in the Daniell,	134
Copper-Plating,	135
Irregularity of the Chemical Action in Daniell's Batteries,	137

CHAPTER IV.

BATTERIES DERIVED FROM THE DANIELL.

Marié Davy's Sulphate-of-Mercury Battery,	140
Weakening of the Sulphate-of-Mercury Battery,	143
Sulphate-of-Mercury Gravity Battery,	146
Trouvé's Reversible Battery,	147
Gaiffe's Battery,	147
Latimer Clark's Standard Battery,	148
Sulphate-of-Lead Battery,	150
Weakening of the Sulphate-of-Lead Battery,	152
Various Salt Batteries,	153

CHAPTER V.

ACID BATTERIES.

	PAGE
Grove's Battery,	154
Chemical Actions in Grove's Battery,	156
Bunsen's Battery, French Model,	158
Bunsen's Battery, German Model,	172
Bunsen's Battery, Faure's Model,	174
Electro-motive Force and Resistance in Nitric-Acid Batteries,	174
Maynooth's Battery,	175
Daniell's Experiments upon the Size and Place of the Electrodes,	176
Chloric-Acid Battery,	177
Chromic-Acid Battery,	177
Various Acid Batteries,	177

CHAPTER VI.

OXIDES IN BATTERIES.

Peroxide-of-Lead Battery,	179
Peroxide-of-Manganese Battery,	180
Leclanché's Battery,	180
Leclanché's Agglomerated Mixture Battery,	189
Clark and Muirhead's Modification of the Leclanché,	193
Electro-motive Force, Polarization,	194
Chemical Action,	194
Weakening of the Leclanché Battery,	196
Practical Durability of the Leclanché Battery,	197

CHAPTER VII.

CHLORIDE BATTERIES.

Chloride-of-Platinum Battery,	200
Chloride-of-Silver Battery,	201
Gaiffe's Battery,	206
Chloride-of-Lead Battery,	208
Perchloride-of-Iron Battery,	208

CHAPTER VIII.

DEPOLARIZING-MIXTURE BATTERIES.

	PAGE
Potassium-Chlorate and Sulphuric-Acid Batteries,	210
Bichromate of-Potassium and Sulphuric-Acid Batteries	211
Chemical Action in the Bichromate Battery,	213
Application to the Telegraph,	216
Gaugain's Experiments,	217
Use in England,	218
Fuller's Battery,	218
Military Batteries,	220
Grenet's Bottle Battery,	222
Trouvé's Battery,	224
Byrne's Pneumatic Battery,	226
Agitation of the Liquid,	228
Camacho's Battery,	231
Delaurier's Battery,	232

PART THIRD.

VARIOUS BATTERIES.

Dry Piles,	235
Identical Electrode Batteries,	237
Unattacked Electrodes in Batteries,	238
Becquerel's Oxygen-Gas Battery,	239
Coke-Consuming Battery,	240
Gas Batteries,	242
Secondary Batteries,	243

TABLES.

Electric Conductibility of Solids,	253
Specific Resistances,	254
Conductibility of Liquids,	255

CONTENTS.

	PAGE
Resistances of Liquids,	256
Dilute Sulphuric Acid,	257
Resistance to Different Liquids,	258
Electro-motive Forces,	259
Remarks upon the preceding tables,	264
Conclusion,	265

PART I.
SINGLE LIQUID BATTERIES.

CHAPTER I.
INTRODUCTION.

A BATTERY, or *pile* as it is sometimes called, is an apparatus arranged to furnish a continued flow of electricity, to which the name of "electric current" is given.

If one should wish to make a complete enumeration, it would be necessary to note:

1. *Hydro-electric* batteries, to the study of which the present work is devoted;

2. *Thermo-electric* batteries, which have as yet received but few applications.

It may be well to state, however, that batteries are not the only apparatus able to produce currents; certain machines produce effects exactly similar.

ORIGIN OF THE NAME OF PILE.

The word *pile*, though not as frequently used as the word *battery*, is, however, more correct.

The invention of *electric piles* is due to Volta, Professor of Natural Philosophy at Pavia, and dates from the year 1800.

One of the first that he constructed was composed of a certain number of discs made of zinc, copper, and cloth *piled* one upon another. In all courses of natural philosophy models of *Volta's pile* are shown, and

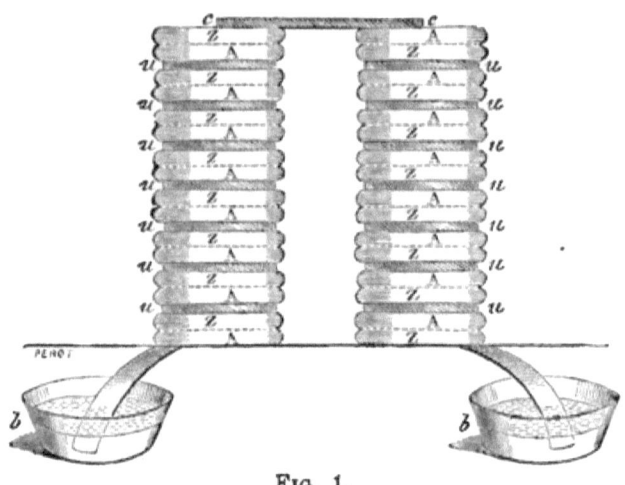

Fig. 1.

Fig. 1 shows the appearance of the instrument called the column-pile, which has to-day but an historical interest; it is a *pile* of discs.*

FIRST IDEA OF THE PILE,

OR BATTERY, AS WE SHALL HEREAFTER CALL IT.

If you immerse a thin plate of commercial zinc into

---

* This figure is a fac-simile of the first cut published of the battery. The original cut is to be found in the "Philosophical Transactions" for 1800.

dilute sulphuric acid, a very lively action takes place; the zinc dissolves, and a considerable quantity of hydrogen is given off. It is indeed this process which is generally employed in the preparation of hydrogen gas.

But if, instead of ordinary zinc, which contains impurities, zinc rendered perfectly pure by distillation be employed, the action takes place very slowly, the bubbles of hydrogen remain attached to the plate of zinc and protect it from further action of the acid.

If a thin plate of platinum, or a platinum wire, be now placed in the same, as soon as the two metals touch at one point the action becomes extremely energetic; the zinc dissolves and hydrogen is given off, but from the platinum and no longer from the zinc.

As soon as the contact of the two metals ceases, all action upon the zinc and all giving off of hydrogen are suspended.

This important experiment, due to De La Rive, throws a great deal of light upon all that follows. It is equally successful when you substitute for the platinum silver, copper, or even iron; it gives the same result when the metals have their point of contact either *in* the liquid or *out* of it.

It permits us to explain the difference in the action of the sulphuric acid upon pure zinc and impure zinc; the heterogeneous particles (of iron or of other metals) found at the surface of commercial zinc play the same part as the platinum. You will observe, in effect, that the hydrogen is only given off within very limited points, and at the end of a certain time the surface becomes rough, which shows that the attack has been more active at some points than at others.

Let us resume the fundamental experiment of De La Rive.

Suppose the two metals to have their point of contact *not in* the liquid but *out* of it, as Fig. 2 represents. The chemical action takes place *in* the liquid, as stated above.

It also takes place if, instead of bringing the two plates of metal into *direct* contact, you put one upon the upper part of the tongue and the other upon the under part.

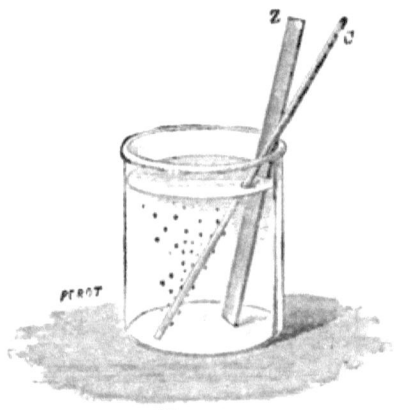

FIG. 2.

You will experience a slight sensation like that of a feeble electric shock, and also a peculiar taste.

If you place upon the dry part of the zinc a strip of paper dipped in iodide of potassium, and then touch this dampened paper with the platinum, a blue spot is immediately produced, which shows that the iodide has been decomposed and iodine set free.

These experiments can also be made if you attach to the zinc and platinum two wires (indeed very long ones may be used), and operate with the two loose ends. If you place one of these in the neighborhood of a freely suspended magnetic needle, you will notice that the needle deviates slightly from its north-south direction as soon as the contact is established between the two loose ends of the wires.

These different observations prove that a singular phenomenon takes place in the two wires, which is the cause of various actions, physiological (upon the tongue), chemical (upon the iodide of potassium), magnetic (upon the needle).

The analogy of these phenomena with those which electric machines with circular glass plates produce, and which were known long before, is easy to comprehend. It is said that an electric current runs over the wire, and one can see from its effects that it is continual.

The two metal plates immersed in the liquid (Fig. 3)

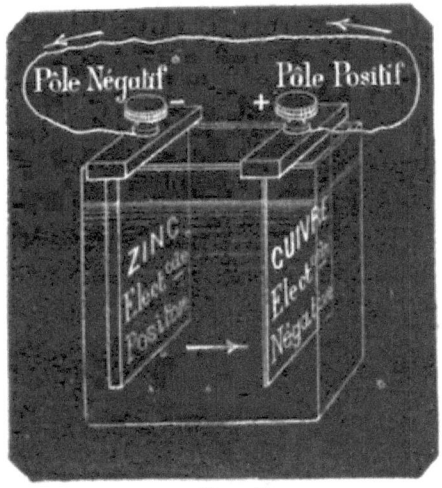

Fig. 3.

are called *electrodes ;* the wires, long or short, attached to electrodes, and which permit the transference to a distance of the effects produced by the battery, are called *rheophores*.

The *rheophores* are generally short, and often end in a longer wire, *c c c c*, to which the name of *conductor* is given.

The name *circuit* of the current is applied to the whole, formed by the battery, the rheophores, and the solid or liquid conductor through which the current passes. In the experiments mentioned above, the tongue and the paper dipped in iodide of potassium formed part of the circuit.

Every apparatus which produces a current is indeed a *battery*. However, the simple apparatus mentioned above (Fig. 3) is, to be more exact, a *cell*, or an *element*, of a battery, and a number of these *cells* grouped together is properly a battery.

It is said that the circuit is *open* when at any point whatever the conductor be disconnected; all the effects of the current then cease and the *current does not circulate*. The current is *closed* when the two parts of the conductor, which were separated, are brought into contact with each other and the current commences to flow.

It is said that a battery is in *short circuit* when the conductor connecting its poles has a null resistance; that is, when it is very short. We will frequently have occasion to use this expression in the course of the present work.

It has thus come to be said that, *in the conductor, the current flows from the positive pole of the battery* (+plate of copper) *to the negative pole* (−plate of zinc); a transference of a peculiar fluid from one to the other of these points is thus implicitly admitted. Let us say, in passing, that this way of looking at things, after having been abandoned in science, shows a tendency towards reacceptance with a few changes, so that the conventional language, which had not been changed, finds itself again in accordance with the theoretical ideas admitted.

The cell formed of the electrodes of zinc and copper immersed in sulphuric acid is more particularly known under the name of Volta; by changing the nature of the liquid and the electrodes, you can obtain an indefinite number of cells which produce the same kind of energy.

PROPERTIES OF AMALGAMATED ZINC.

We have shown, in that which precedes, how differently the pure zinc and the ordinary commercial zinc act in the voltaic cells.

The result is that when pure zinc is employed there is no local current at its surface, and that the electricity which is produced passes entirely into the circuit between the poles, and also that the hydrogen is given off from the copper.

If, on the other hand, impure or commercial zinc be employed, the giving off of hydrogen takes place, for the most part, upon its surface; there is reason to conclude, from this, that a very large proportion of the chemical action is lost for the production of the electric current.

Thus, in the construction of batteries, the use of pure zinc presents very important advantages; but the price of this material is almost fabulous, and it can almost be called a curiosity of the laboratory.

Happily, a very simple artifice has been discovered, by which the properties of pure zinc may be given to commercial zinc. It suffices to amalgamate it—that is, to spread mercury over its surface in such a manner as to form a layer of amalgam of zinc. This amalgam is an alloy, or, in other words, a combination of zinc and mercury.

The experiment shows that *the* amalgamated zinc immersed in sulphuric acid diluted with water, is scarcely attacked, and if it be employed as the positive electrode of a voltaic cell, it occasions no local actions; the giving off of hydrogen takes place entirely upon the negative electrode, of copper or platinum.

In short, amalgamated zinc presents, for use in batteries, the same advantages as the chemically pure zinc, and with a few exceptions zinc should always be amalgamated.

INCONSTANCY OF SIMPLE BATTERIES.

All the cells of which we have spoken, formed of two electrodes immersed in a liquid, present an immense drawback; namely, their action decreases very rapidly from the beginning of the action.

The causes of this decrease are twofold, which we will analyze summarily here.

The first is the loss of acid from the dilution. It can be easily understood that water acidulated in the proportion of 1 to 100 will act less energetically than water acidulated in the proportion of 1 to 10. This cause of the weakening of the battery is not felt until the expiration of a certain time, and it is easily avoided by adding, from time to time, acid to the dilution.

The second is the deposit of hydrogen upon the copper. If the current be interrupted during a length of time sufficient for the freeing of the hydrogen, it will be seen, as soon as the current is again closed, that the intensity assumes its original worth; it suffices indeed to agitate the plate of copper in order to cause the gas to free itself and to give to the current its initial intensity.

Constant batteries are those in which this second cause of weakening, called *polarization of the electrode*, is removed. The presence of the hydrogen upon the electrode opposes a double resistance to the passage of the current, a *passive resistance* and an *active resistance;* it is the latter that is properly called *polarization of the electrode.* To *depolarize the electrode*, is to provide against these resistances by suppressing the freeing of hydrogen.

It is very important to comprehend perfectly everything pertaining to this question; therein lies the whole difficulty concerning the improvement and perfecting of batteries. We will return to it in the course of our exposition.

Various reasons have combined to designate the *positive electrode* as that one which represents the negative pole of the cell (zinc, in Volta's battery), and *negative electrode* as that one which represents the positive pole (copper or platinum, in the cells which have occupied us up to the present).

One of these reasons has been indicated above, which is that the current enters the liquid of the battery by the negative pole, and goes out by the positive; in other words, the *positive electrode* is that by which the electricity enters the cell.

However excellent may be this reason and those which we will give further on for the choice of these denominations, it is not to be denied that they are difficult to employ. In reality, this difficulty may be avoided by speaking of the positive pole and negative pole, when you want to designate the corresponding electrodes; that is what the majority of practical men do. But if you wish to employ absolutely correct and scientific terms, take great care not to apply them wrongly, as you will only arrive at confusion by an awkward research for precision in the language.

We find in the excellent book, "Daniell's Introduction to Chemical Philosophy," another denomination which ought to be employed more frequently than it is, because it presents the expression of a *fact* and does not depend upon theoretical ideas, which are always open to discussion.

He calls the *generating electrode* that one which plays a part in the chemical action; it is the zinc in the cell that we have considered.

He calls the *conducting electrode* that one which is not attacked, and which serves, however, to complete the cell.

The first can also be called *soluble electrode*.

BATTERY CELLS JOINED IN INTENSITY.

We have described above the most simple cell that can be prepared, composed of two electrodes of copper and zinc immersed in acidulated water.

The cell of Volta's column-battery does not differ essentially from this one; it is composed of two discs, one of copper and the other of zinc, separated by a circular piece of cloth saturated with acidulated water.

Two "rheophores," or copper wires, are soldered to these two discs and conduct the current to apparatus upon which it is to act.

But as we have summarily indicated from the commencement, Volta placed upon this first group of three discs (zinc, wet cloth, copper) a second group entirely identical and disposed in the same order; then a third, a fourth, and so on.

These discs, in various quantities, the one at the top being of copper and the one at the bottom of zinc, constitute the battery of Volta.

Volta discovered, by delicate means, that the force of the current increased as the number of cells was augmented, and made one of the most brilliant inventions of modern times.

He thus showed that it was possible to add one source of electricity to another and to a third in such a manner as to obtain a multiple source of an indefinitely increasing power.

Although three quarters of a century have passed since this discovery, it is not certain whether all of its resources have been exhausted, and it is probable that unlooked-for consequences may yet be brought to light.

It is remarkable that he made at the same time an invention and a discovery; he invented an apparatus, a machine, an implement, which has received and will receive many applications: at the same time he discovered one of the most fruitful principles of physics, to which he opened a new road.

If you should wish to show the increase of force of a battery with the number of cells or groups of three discs, the most simple means consists in causing the current to act upon a galvanometer or detector. The deflection of the galvanometric needle would be seen to increase in proportion to the number of cells; that is indeed a fundamental truth, verified by experiments at every moment.

The copper electrode of the cell of Volta (Fig. 3) is the *positive pole*, the zinc electrode is the *negative pole* of the cell.

When the cells are piled up or joined in intensity as in Volta's battery, the *positive pole* of the battery is that of the last cell, and the *negative pole* is that of the first cell.

In order to give an exact definition of the *positive pole of a battery*, or of a cell of a battery, it is necessary to say that it is that one whence the current starts circulating in the exterior conductor, and that the *negative pole* is that one towards which this same current flows, as shown by the arrows, Fig. 3.

To be complete, it must be added how the direction of the current may be recognized. The wire through which the current flows being placed directly over a freely suspended magnetized needle, causes the north pole of the needle to deflect towards the west, when the current flows from south to north.

These preliminaries being established, we may enter upon the description of the principal arrangements of Volta's Battery.

CHAPTER II.

THE VOLTAIC BATTERY AND ITS DERIVATIVES.

COLUMN BATTERY.

We have described this battery in the preceding pages. We add that it may be vastly improved upon by soldering the disc of copper of each cell to the disc of zinc of the following cell; all faulty contacts of the metal plates are thus avoided.

The discs of cloth should be smaller than the metal discs. It is noticed, however, after a short time that the weight of this column squeezes out the liquid from the cloth; this liquid runs out over the edges of the discs and soon disappears, so that the battery rapidly weakens, and after a certain time produces no effect whatever.

VOLTA'S "COURONNE DE TASSES."

It is generally admitted that the column battery was the first one that Volta arranged. This is, however, not correct; the "couronne de tasses" was the first; and according to us is much preferable. A series of glasses or cups were placed in a circle, forming a kind of a crown; plates of copper and zinc were so arranged that, being connected at the top, the plate of zinc was placed in one cup and the plate of copper in the next.

This battery is truly the model of all those existing

to-day, and will be our model for reference in the description of others.

It is interesting to note that Volta did not think of the column-battery until afterwards, and then it was with a view to produce an instrument that might be easily transported into hospitals for medical purposes.

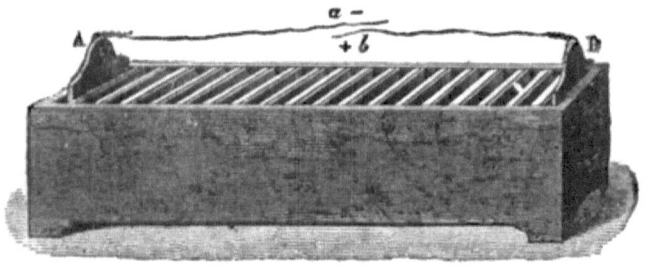

Fig. 4.

CRUIKSHANK'S BATTERY.

This battery is composed of a wooden trough, internally coated with marine glue and divided into cells separated by metallic partitions; these partitions are composed of two thin plates, one of zinc and the other of copper, soldered together. They are arranged in such a manner as to have all the plates of zinc on the same side and all the plates of copper on the other. The cells thus disposed in the wooden trough are nearly filled with acidulated water, and if they are water-tight the battery thus constructed is very satisfactory.

It is not necessary to enumerate the inconveniences of Cruikshank's battery, which is no longer in use; we would only point out the impossibility of changing the plates of zinc when they have been partially destroyed by the action of the acid.

WOLLASTON'S BATTERY.

The difficulty mentioned above is not to be found in the battery combined by Wollaston.

The pairs of metallic plates (zinc and copper) are attached to a cross-bar of wood, which allows them to be

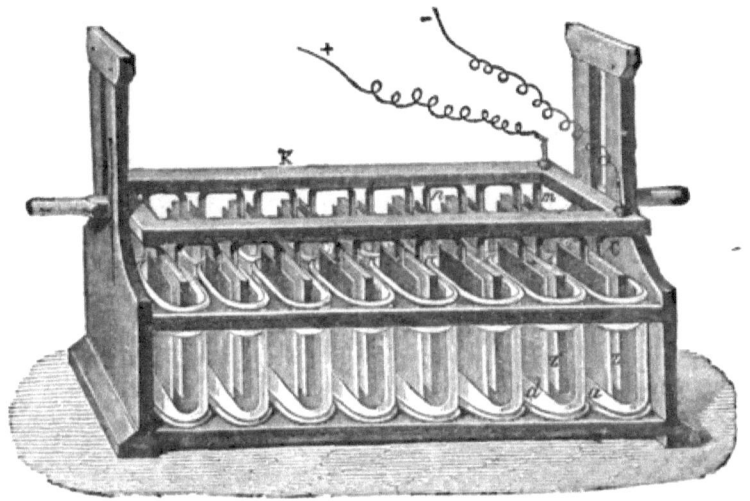

FIG. 5.

lifted out or immersed all at the same time in the glass vessels.

This arrangement is excellent, and is still employed very frequently.

Wollaston made another change in the combinations adopted before his time: he placed the plate of zinc in the centre and surrounded it with a thin sheet of copper, thus giving to the negative element a surface double that of the zinc. The reasons of this disposition are several, upon which we will remark:

1. When two plates are immersed in a liquid, the

two sides facing each other alone combine in producing the current; the other sides could be covered with an insulating coating without notably diminishing the current. In Wollaston's disposition, the two sides of the zinc become active.

To this it might be opposed that an inverse disposition would present the same advantages, and that a plate of copper might be placed between the two plates of zinc

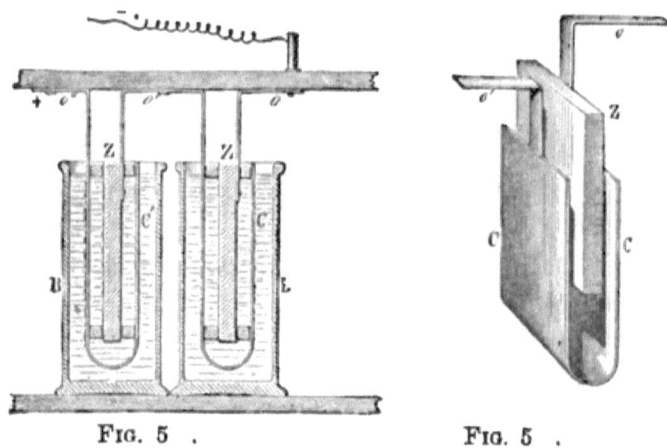

Fig. 5 .   Fig. 5 .

so as to make use of the two sides of the copper and only the half of the surface of the zinc. But as the zinc is subject to local action or waste, its size should be reduced to just that amount which is requisite to maintain the current required. There is, on the other hand, no disadvantage whatever in increasing the immerged surface of the copper, as this metal is not attacked by the dilute sulphuric acid.

There is, we repeat, an advantage in reducing the surface of the zinc as much as possible; for when the battery is not in use and the electrodes, however, remain immerged in the liquid, the attack upon the zinc continues, although with less intensity, and this dissolving of the zinc is pure loss. As this waste is evidently in proportion to the immerged surface, it is best to have the least possible surface of zinc; or better, to have *no* part of that surface which may be useless for the producing of the current.

2. We have stated above that hydrogen is given off from the positive electrode, and that this *polarization of the electrode* was a cause of weakening of the current of the battery.

If the hydrogen would free itself as it is generated, the production of the electricity would not be perceptibly diminished; but it does not free itself—that is, not wholly—and what remains, tends to reduce considerably the intensity of the current. It is evident that the smaller the surface the more rapidly a certain quantity of hydrogen, being produced upon the positive electrode, will act; in other words, the larger the surface to be polarized, the more slowly the effect of the polarization will be felt.

This is the second reason given for the disposition of Wollaston, in which the surface of the zinc is entirely surrounded by the surface of the copper. We will return to this subject farther on, in speaking of the action of the air upon batteries.

SPIRAL BATTERY.

The two electrodes of this battery are rolled parallel to each other in the form of a helix, and separated by a tissue of osier; in the centre is a wooden handle to which the whole apparatus is attached, and by which it may be lifted. It is immersed in a bucket of acidulated liquid, and thus you have electrodes with very large surfaces separated by a very short distance; the interior resistance of the battery is consequently much reduced, and the quantity of electricity produced very considerable.

This battery presents some of the advantages of that of Wollaston, inasmuch as both surfaces of the zinc are

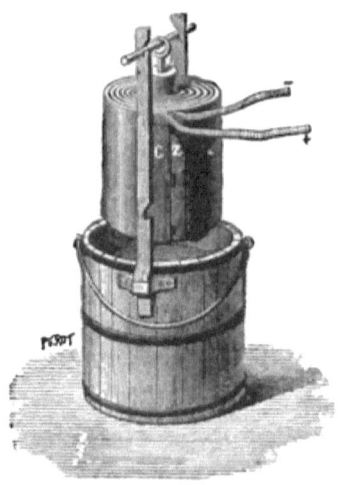

Fig. 6.

used; on the other hand, both surfaces of the copper are also used.

Cells of this description may be joined in intensity as those of an ordinary battery; but they were more frequently used separately.

The spiral battery has indeed been entirely abandoned since the inventions of Grove, and Bunsen of Poggendorff (with bichromate of potash).

MUNCKE'S BATTERY.

Wollaston's battery being cumbersome and unwieldy, Muncke, Young, the illustrious Faraday, and others imagined various ingenious arrangements for joining a large number of cells in a small volume.

In Muncke's arrangement, the parts where the electrodes of zinc and copper are soldered together are placed vertically; they are divided into two series, the one fitting in the other as Fig. 7 represents.

This battery, and the one arranged by Faraday, which

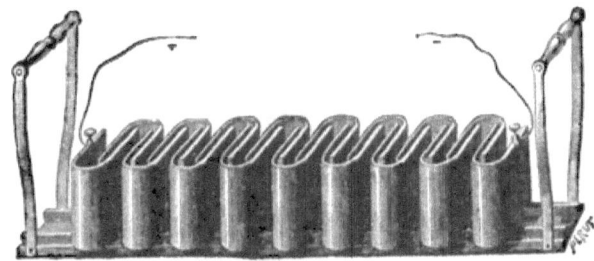

FIG. 7.

differs from it very slightly, were employed for several years in laboratories, as the whole battery could be immerged in one trough, which was very convenient. They are completely put aside to-day.

SAND BATTERY.

This battery is composed of a trough made of teak, divided into cells by partitions of slate or of wood; to make it water-tight it is coated internally with marine glue. A plate of amalgamated zinc placed in one cell is joined to a plate of copper in the adjoining cell, and resting, at their point of contact, upon the partition; the cells are then filled with sand saturated with acidulated water.

This battery is to-day abandoned, but it presented

many practical advantages. It was used for a long time in the telegraph service, needing no attention for several weeks at a time, and was much more easily moved from one place to another, than batteries wherein the liquid might be spilled when carried about.

NATURE OF THE CHEMICAL ACTION IN VOLTA'S BATTERY.

All the batteries that we have just described differ only in their arrangement from that of Volta's; in every one we find the zinc, the copper, and the water acidulated with sulphuric acid.

The chemical action is very simple. Under the influence of the water and sulphuric acid, the zinc becomes oxydized; the oxide of zinc uniting with the acid produces sulphate of zinc, and the hydrogen of the water is given off upon the electrode of copper.

Thus, on one hand we have the dissolving of a metal (zinc) in the liquid, and on the other the freeing of a metal (hydrogen) which is extracted from the liquid of the battery. Hydrogen, although gaseous, is considered by chemists as a metal.

It will be seen, as we advance, that the action is the same in nearly all batteries: dissolving of one metal, freeing of another. On account of its importance in nature and in chemistry, hydrogen will, of all metals with which we will have to do, be the one the most frequently freed under the influence of the battery. Far from presenting an exception to the preceding rule, this is a confirmation and a capital example.

All our readers know that when they prepare hydrogen gas for use in laboratories, they place small bits of zinc in an appropriate jar with acidulated water.

Since there is an attack upon the zinc without the intervention of any other metal, it can be seen that in all the forms of Volta's battery hydrogen gas will be given off and the zinc will be dissolved without closing the circuit; that is, without the production of electricity by the battery. This is one of the greatest faults of these batteries; they are consumed without doing any useful work, like a horse who stands in the stable and eats without working.

In will be seen, in that which follows, that nearly all batteries present this same difficulty; there are, however, a few exceptions, upon which we will bestow particular attention.

The hydrogen given off under the chemical action of the battery appears upon the negative electrode of copper; it is seen in the form of bubbles which rise and leave the liquid more or less rapidly. But in addition to these visible bubbles, there is a large quantity of gas deposited upon the surface of the electrodes and which is not seen. This invisible layer of gas is of great importance in the study of batteries, and produces, as we have already stated, the *polarization of the electrode*. We are thus brought again to speak of this phenomenon, so important in the study of batteries, and of which it is the most delicate point. We have taken the opportunity of showing how this injurious action may be overcome, and how to obtain a partial *depolarization*.

ACTION OF THE AIR UPON BATTERIES.

The air acts very favorably upon batteries on account of the oxygen it contains.

At the time of the discovery of the battery, it was noticed that ordinary cells exposed to the air absorbed the oxygen, and that the current had a tendency to stop when there remained nothing but nitrogen. But observation shows that the effect is due, not to the action of the oxygen upon the zinc, but to a *depolarization* of the other electrode. In the cells of Volta and Wollaston, the action of the oxygen is experimentally demonstrated.

It will be noticed that this depolarizing action is greater in Wollaston's battery, which is a new reason explaining the advantages of giving to the negative or conducting electrode a considerably larger surface than that of the generating electrode.*

---

* These remarks are only correct when concerning single-liquid batteries. There is no action of air in batteries totally depolarized, like that of Daniell.

CHAPTER III.

GENERAL REMARKS UPON BATTERIES.

IDEAS UPON ELECTRIC RESISTANCE.

WE have said that the most simple way of showing the passage of electric currents in a conducting body is to bring its force to bear upon a magnetic needle.

Let us suppose that the conductor of a galvanometer, or of a simple detector, be inserted in the circuit of the current of a battery, and that the deflection of the needle be 25°, for instance. Now if the circuit be lengthened by the addition of a wire, the deflection will be seen to diminish to 15°, and if the circuit be made still longer, the deflection of the needle will not exceed 10°. From this experiment several conclusions may be drawn:

1. The intensity of the current is less in the second instance than in the first, and less in the third than in the second.

2. The influence of the additional wire being only passive, the reduction of the intensity of the current is due not to the decrease of the generating force, but to the increase of the resistance.

These experiments give an idea of the *resistance* that conducting bodies offer to the passage of currents; and they also demonstrate that the resistance of a conductor increases with its length.

Very exact and oft-repeated measurements have proved

that the resistance of a conductor is in proportion to its length and in inverse proportion to its sectional area.

We will not dwell upon the demonstration of these laws, which are found in all works upon physics. It suffices for practical men to know the formulæ of these rules which are constantly being applied.

GENERAL REMARKS UPON ELECTRO-MOTIVE FORCE AND RESISTANCE.

In all machines in motion is seen a power or cause of movement; and there are also resistances which tend more or less to slacken this movement or to stop it altogether. Let us take, for instance, a windmill. The large arms, under the pressure of the wind, cause the millstones to turn which crush the grain. In the working of the mill we see first a power, the wind, which produces the movement.

Then there is a resistance offered by the grinding; this resistance moderates the pace of the arms, and if the wind falls it stops them entirely.

At first sight there are two mechanical elements apparent: the power or cause of movement, or motive force; and the resistance, or work.

A careful examination will show, however, that the resistance is complex, and that that offered by useful work, as the grinding, should be distinguished from that which is the result of the friction of the different parts of the machine in motion, and of certain secondary phenomena. All practical men know that a badly oiled rubber is sufficient to slacken the movement of a machine, or even to stop it; all know the importance of friction in the different parts of the machine, and of the stiffness of the belts and ropes. These inevitable causes of the slackening, which absorb a part of the motive power at the cost of the useful work desired, are called passive resistances. Every one knows that these resistances should be diminished as much as possible, and that they cannot be totally suppressed.

Attention should be called to the fact that in many cases no useful work is done, and that there then remain only passive resistances. If the miller takes away his millstones and still permits the mill to turn, it is evident that there remain only those passive resistances (friction and others) which are produced by the machinery remaining in motion. If all the machines of a large factory be disconnected from the motion-giving steam-engine and the engine continues to turn, there will only be present the motive force furnished by the engine itself and the passive resistances existing in the engine, in the shafts, and in the different agents of the transference of the movement which are still in motion. If now the steam-engine runs entirely alone, not being connected with any shaft or any piece of machinery outside of itself, we have not only the example of a system in which there are force and passive resistances, but also that particular instance where these passive resistances are inherent to the force-giving machine and inseparable from the production of that force.

In a circuit through which an electric current flows, the same terms are to be found: first, a force residing in the battery and which is called *electro-motive force;* next, the work; and finally the passive resistances. The work may be found in the movement of the clapper-spring of an electric bell; it may be in the movement of a telegraph instrument placed at a great distance from the battery; it may be in the movement of an electro-motor or an electro-magnetic machine which lifts a weight; it may be in a chemical decomposition produced by the passage of a current in the production of heat and consequently of light in a voltaic arc, etc. etc.

Passive resistances are the results of the circulation of the current in the different parts of the circuit. We have explained how their existence may be ascertained, and we have designated them by this one word *resistance.*

If the current produces no real work—that is, if the circuit is composed solely of conductors without the interposition of any apparatus which puts the current to any use—the resistance is entirely passive. These considerations explain and justify the use of the word *resistance* applied to that property of reducing the intensity of the electric current which the conductors possess, and which we have demonstrated in the preceding chapter.

ELECTRO-MOTIVE FORCE.

The cause which produces the electric current we have called *electro-motive force.* Before going farther we will show several experiments, which will render the ideas upon this force more precise.

Let us take a battery cell (Fig. 3—zinc, copper, and water acidulated with sulphuric acid) and cause the current which it produces to act upon a galvanometer, and we will see that the needle is deflected, for instance, towards the right. If we change the communications of the battery with the galvanometer, the direction of the needle's deflection will be altered, which shows that the direction of the current in the galvanometer has been changed.

GENERAL REMARKS UPON BATTERIES. 27

But let us consider the first conditions: the needle is deflected towards the right.

Let us now take a second battery cell, differing in no way from the first, and insert it in the circuit. If the negative pole of the second be attached to the positive pole of the first, the two currents flow in the same direction and join each other; the intensity of the resulting current is increased, and consequently the deflection

FIG. 8.

of the needle is greater. In these conditions the two battery cells are *joined in intensity* (Fig. 8); they form a battery of two cells. A battery of any number of cells could thus be formed as we have stated above, but that is not the point upon which we wish to insist; we only desire to recall the expression, *battery cells joined in intensity*, and to determine its exact meaning.

Suppose now that the second cell be inserted in the circuit of the first; by uniting the positive pole to the positive pole, and the negative to the negative, in such a manner as to have two poles of the same name ending at the galvanometer (Fig. 9), the needle will remain stationary. This is not to be wondered at, if it be remembered that the two cells tend to produce equal currents in opposite directions. It is quite natural that these currents balance each other, and that there is no movement either in one direction or the other. It is

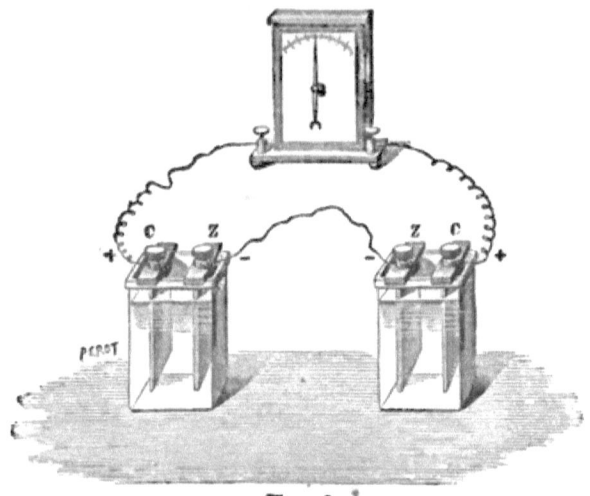

Fig. 9.

said in this case that the two battery cells are *opposed* to each other, or are *joined in opposition*.

We have assumed, in the preceding, that the opposed cells were of equal dimensions. Each one acting alone would produce the same deflection of the needle, one towards the right and the other towards the left; both acting simultaneously in opposite directions cause no deflection whatever: which is quite natural and easily understood.

Let us now vary the experiment, and place in the

same circuit (Fig. 10) a small voltaic cell in opposition to a larger one of the same nature; the needle will remain stationary, thus showing that there is no current. This result will appear very strange to the uninitiated reader, and deserves to be dwelt upon. If they are made to act separately, they cause the needle to deflect, one towards the right, the other towards the left. The current furnished by the larger one is more intense than the current

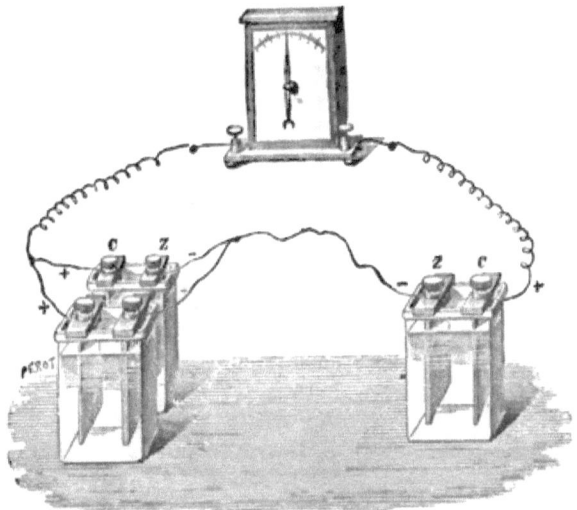

FIG. 10.

produced by the smaller one, as the deflections of the needle show. But if these two cells be opposed to each other, the effect of one is counterbalanced by the effect of the other, and no current flows through the circuit. The conclusion of this capital experiment is that *the electro-motive force of battery cells does not depend upon their dimensions.*

The above experiment may be slightly modified. When cells of equal dimensions are opposed to each other, there is no deflection of the galvanometric needle. You may lift up the zinc or the copper of one of the cells, or even the zinc and copper together of one of the cells; you may, in a word, increase or diminish the immersed part of the electrodes of one of the cells, and still there will be no deflection of the needle, and the electro-motive forces remain equal.

To elucidate still further this subject, we will present a few more experiments.

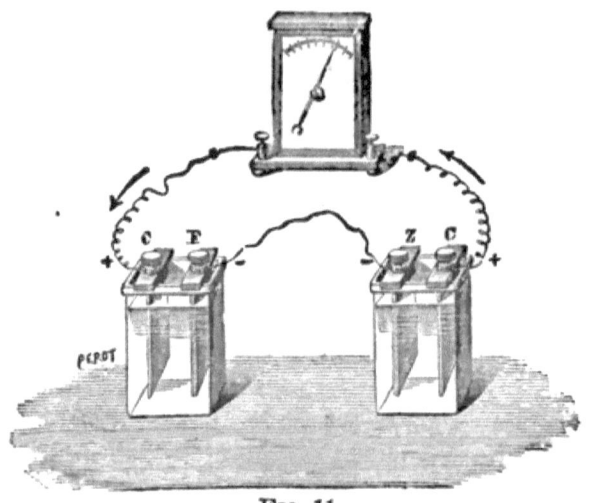

FIG. 11.

Place two cells in opposition to each other, the one similar to those of which we have spoken (zinc, copper, and dilute sulphuric acid), and the other differing but slightly in appearance (iron, copper, and dilute sulphuric acid). The difference is the substitution in the second of iron for zinc. A first trial will show that the copper is the positive pole in the second cell as in the first; that is, the current flows from the copper to the iron in the second, as it does from the copper to the zinc in the first. Place them now in the same circuit, in opposition to each other—that is, join the two zinc poles and connect the other two with the wires of a galvanometer (Fig. 11); the needle will be seen to deflect in the same direction as if the voltaic cell were acting alone, although the deflection is less. We have a right to conclude from this that the first cell has a greater electro-motive force than the second, and that *the substitution of iron for zinc in Volta's battery would be detrimental.*

In this experiment we have supposed the two cells to be of equal dimensions, and that the electrode of iron was the same size as that of zinc. We can now modify these dimensions. Let us suppose, for instance, that a very small voltaic cell be placed in opposition to a very large cell (iron, copper, and acid).

The direction of the deflection will be the same as in the preceding experiment; that is, the electro-motive force of the smaller cell is greater than that of the larger one. This new experiment proves again, and more clearly than ever, that *the electro-motive force of battery cells does not depend upon their dimensions, but upon the materials used in their composition.*

MEASUREMENT OF ELECTRO-MOTIVE FORCES.

It has been seen how, by means of an ordinary galvanometer, the electro-motive forces of different batteries may be compared. The method that we have used is called *method of opposition*, because it consists in opposing equal or unequal forces to each other.

It can be easily understood how the electro-motive forces of different cells may thus be measured and tables of these forces made out.

Let us take two batteries, A and B, of unequal electro-motive forces. A first experiment will show us, for instance, that A is stronger than B. By opposing A to 2B we find that 2B is stronger than A. Let us now oppose 2A to 3B, and if there is no deflection of the galvano-metric needle we may conclude that twice the electro-motive force of A is equal to three times that of B, or that $A = \frac{3}{2} B$, or, finally, that $A = 1\frac{1}{2} B$.

It is seen that this method is general; it may be varied advantageously in different ways. We will not insist upon it any longer, as we only wished to show the possibilities of these measurements and not the way to obtain them.

INTERNAL RESISTANCE OF THE BATTERY.

It has been seen from the foregoing that the conductors outside of the battery offer a certain resistance to the electric movement, or, in other words, a resistance to the passage of the current.

We will now show by several simple experiments that the battery itself offers a resistance to the current it produces.

The elementary battery (Fig. 3) is made to act upon a galvanometer. Observe the deflection. Lift up gradually one of the electrodes, and as the immersed surface becomes less the deflection diminishes.

The result shows a decrease in the intensity of the current. As our former experiments have shown, however, that the electro-motive force does not vary under these circumstances, and that the other parts of the circuit do not change, we are justified in saying that the resistance of the battery has increased.

The result would be the same if the two electrodes were lifted at the same time.

The experiment may be made by separating the two electrodes from each other, still having the same extent of surface immersed. It is perhaps in this manner that the experiment is made the most clear. In these experiments the intensity of the current is seen to change with the distance that separates the two electrodes in the trough of liquid and with the section of the trough. It may be concluded that batteries have an internal resistance in themselves, and that *the resistance increases with the distance between the electrodes in the liquid, and diminishes when the immersed surfaces are increased.*

If the battery be considered as a force-producing machine, it is not to be wondered at that it at the same time produces force and offers a resistance to that force. This condition is common to all machines; a part of the force they produce is absorbed by those passive resistances resulting from the action of the different parts of the machine. In a steam-engine, for instance, the friction of the steam in the pipes, the friction of the piston in the cylinder, etc. etc., cannot be avoided.

This resistance of the battery has to be taken into account in nearly all cases for the explanation of phenomena and for the calculation of results.

It can be seen that of two batteries in which the electrodes are of unequal dimensions, the distance between them being equal in each, the one having the larger electrodes offers less resistance than the other; and it can be said in general that large cells, when compared with small ones, offer less resistance, because the increase of surface of the electrodes is greater than the increase of the distance between them.

The resistance of the batteries varies with the nature of the liquids in which the electrodes are immersed. It can be easily understood that all liquids have not the same specific power of resistance. The conductivity of dilute sulphuric acid varies with the proportions of water and acid mixed, and the greatest conductivity is found in a mixture of 29 parts of sulphuric acid (HSO^4) for 71 parts of water. It has been observed that it is this mixture which, in an apparatus for the production of hydrogen, attacks the zinc the most energetically.

These reasons would lead to the use of this mixture in preference to all others in Volta's battery, and indeed in all others in which dilute sulphuric acid is used; but this mixture, being that of about one part of acid (HSO_4) for two parts of water, is not used in the practice, as it would be too dangerous to handle, and as it is also rather costly; therefore the mixture of ten or twelve parts of acid for one hundred parts of water is adopted.

It is understood that as soon as a battery is put into working order and the chemical action takes place, the composition of the liquid changes, and consequently the resistance.

We will return more than once to this important point.

VARIOUS WAYS OF JOINING VOLTAIC CELLS.

We have seen (Fig. 9) how two battery cells of the same kind may be placed in opposition to each other in such a manner as to counterbalance each other. Let us

GENERAL REMARKS UPON BATTERIES.   35

now take away the galvanometer that we had placed in the circuit of these cells and we will still have two cells joined in opposition.

Let us consider the two cells thus joined. If the galvanometer be put into communication, on one hand with the wire connecting the two positive poles, and on the other hand with the wires connecting the two negative poles, the passage of a very strong current will be observed. The currents of the two cells, which were at first

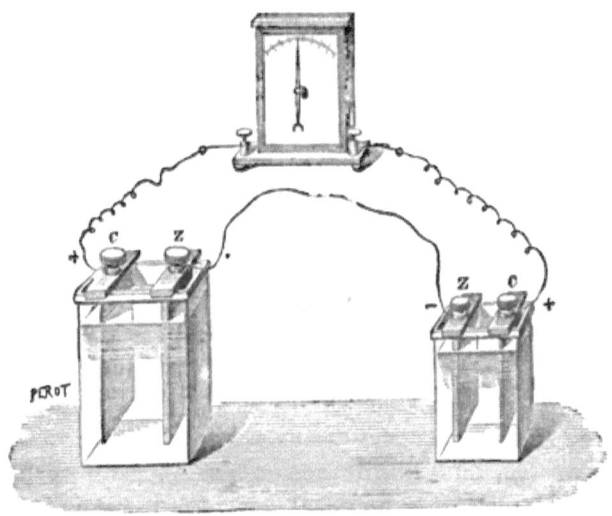

Fig. 12.

opposed to each other, now flow together in the galvanometer. The two battery cells are then said to be joined in quantity.

The metallic piece which connects the two zinc poles may be considered as the negative pole common to both cells, and the other as the positive pole common to both cells.

It may be observed that the two cells ought to produce the same effects as a single one, in which the electrodes would have a double surface, while the distance between them would remain the same.

The internal resistance offered by the two cells is only half of that offered by each one alone, while the electromotive force remains the same. This may be demonstrated by placing a third cell of the same size and kind in opposition to these two cells joined in quantity, Fig. 12. The galvanometric needle does not deflect, which shows once more that the electro-motive force does not depend upon the size of the electrodes, but solely upon their nature.

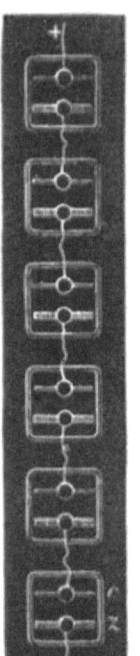

Fig. 13.

There is, finally, a third way of joining these two cells; namely, joining them in *intensity*, of which we have already spoken. This manner consists in uniting the positive pole of one of the cells to the negative pole of the other. In this arrangement the electromotive force of the two taken together is double that of each separately; the resistance is also double.

These different ways of joining battery cells may be applied to any number of cells. Let us take, for instance, six cells and join them in intensity, Fig. 13. If the electromotive force of one cell be symbolized by E, and its resistance by R, it is evident that a battery of six cells joined in intensity will have an electro-motive force equal to 6E, and a resistance equal to 6R.

If all be joined in quantity, Fig. 14, the electro-motive force of the battery will be E, and the resistance $\dfrac{R}{6}$.

GENERAL REMARKS UPON BATTERIES. 37

If they be joined by twos in intensity and by threes in quantity, Fig. 15, the electro-motive force will be 2E, and the resistance $\tfrac{2}{3}R$.

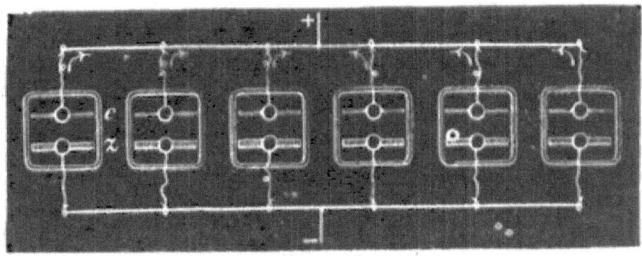

Fig. 14.

They may, finally, be joined by threes in intensity and by twos in quantity, Fig. 16; the electro-motive force will be 3E, and the resistance $\tfrac{3}{2}R$.

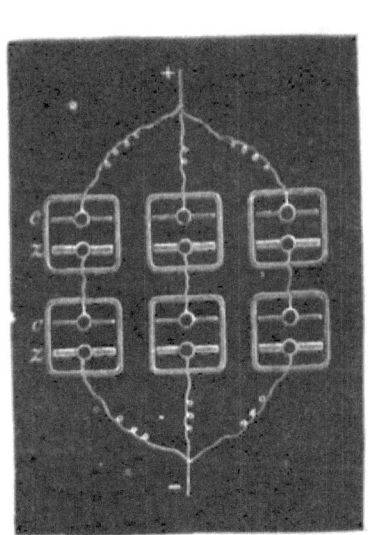

Fig. 15.

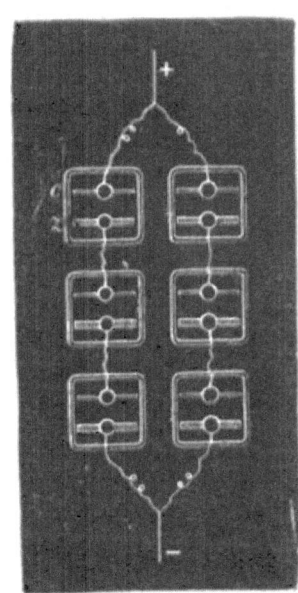

Fig. 16.

As long as, in this last combination, there is no connection with any outside circuit, the three cells on the right are in opposition to the three on the left.

It is not necessary for us to insist longer upon this subject, or to make calculations which are indeed very simple, to make the reader understand that, with a sufficient number of cells, a battery may be obtained whose electro-motive force will be as great, and whose resistance will be as little, as can be desired.

In most applications, and notably in the electric telegraph, there is only the need of increasing the electro-motive force, and very little attention is paid to the resistance.

In certain instances, however, too great a resistance would be very detrimental; it is then that the cells may be joined in quantity. In practice, large cells having a very slight internal resistance are employed.

VOLTAMETER.

Before proceeding with the study of batteries, it would be well to stop and examine some of the effects they produce. Of all the chemical actions that can be brought about by means of electric currents, the decomposition of water is the most striking. It is done in an apparatus called voltameter, and is represented in Fig. 17.

Two wires or plates of platinum are placed parallel to each other in a jar containing dilute sulphuric acid. These two electrodes pass through the bottom of the jar and are attached to binding screws, or terminals, to which the wires of a battery are fastened.

If a sufficiently energetic current be made to pass in this apparatus, bubbles of gas will be seen to free themselves from the surface of the electrodes. If these gases be collected in proper gas-measuring jars, oxygen will be found in one and hydrogen in the other. If they be collected together in a single jar, they will be found to be sensibly in those proportions whose combination produces water. We say sensibly, for the proportion is nearly always altered by complicated disturbing actions, upon which we cannot here enlarge.

FIG. 17.

The electrode by which the current enters the apparatus is called *positive electrode of the voltameter;* it is that which is connected with the positive pole, or, in other words, with the negative electrode of the battery which furnishes the current.

The negative electrode of the voltameter is connected with the negative pole, or positive electrode or generating electrode of the battery.

The *oxygen* which appears upon the positive electrode of the voltameter is termed *electro-negative;* the *hydrogen* which is seen at the surface of the negative electrode of the voltameter is termed *electro-positive.*

These denominations may embarrass beginners. In order to employ them correctly the key is needed, and this may be found in the old theoretical ideas upon the two electric fluids, the one positive and the other negative. There is, at each point in a circuit through which a current flows, a reuniting of positive and negative electricity; the negative electricity of the oxygen is attracted by the positive electricity of the positive electrode, and so on. This circuit is considered as a chain, in which one end of each link is positive and the other negative.

The theoretical ideas have changed, but the expressions have remained, the alteration of which would only involve difficulties, because they are not in disagreement with the new scientific views. We will not enter into the details of this demonstration, but will return to the exact application of these terms, in order to spare the reader the annoyance of certain errors to which he may be exposed.

In general, every liquid decomposed by the passage of an electric current is called an *electrolyte*, and it is said to be *electrolyzed* as long as the electric action continues. Faraday has established, by numerous experiments, the laws of *definite electrolysis*. We cannot enlarge upon this delicate subject. We will only say that, if two or three cells joined in intensity produce a current used to electrolyze water, for instance, for each chemical equivalent of hydrogen set free in the voltameter there will be an equivalent of zinc dissolved in each cell of the battery. The law of Faraday may be said to be the equivalence of chemical work in all parts of the circuit.

If the experiment be made with six cells, instead of with three as indicated above, the quantity of hydrogen set free in one minute is much greater. An idea of the *quantity of electricity* is thus obtained, and it can be understood how the instrument called voltameter permits one to measure this quantity. It owes its name to Faraday, who was perfectly justified in so calling it, as it is in truth an instrument of measurement. The same cannot be said of the galvanometer, which it would be better to call galvanoscope; for in general it does not measure the intensity of the current which passes through it, and it is only by means of complicated contrivances that any measurements can be obtained from its indications.

Unhappily the voltameter is not convenient for use. In many cases it gives no indications, and in others produces false results, on account of the resistance which it introduces into the circuit. It presents other causes of error, as will be seen in the following pages.

SECONDARY CURRENTS.

POLARIZED ELECTRODES.

If the voltameter be submitted for a short time to the action of a current, its electrodes acquire remarkable properties, which may be recognized in the following manner:

Detach the wires connecting the voltameter to the battery, and then connect the voltameter with a galvanometer; the galvanometric needle will be seen to deflect, thus making manifest the passage of a current furnished by the voltameter. The direction of the current is such as to show that that which was the negative electrode of the voltameter in the experiment with the battery has become, in the experiment with the galvanometer, the positive pole of this new source of electricity. In other words, the current flows in one direction in the first case, and in the opposite direction in the second case. It may be said that the voltameter has been charged with a part of the current of the battery, and that the voltameter returns this current in the contrary direction.

It has been said that the *electrodes are polarized*, which is indeed true; for they have been rendered capable of acting as poles. This is the origin of the expression *polarization of the electrodes* which we have already used, and which we will frequently have occasion to employ.

The current furnished by the polarized electrodes of the voltameter in the conditions indicated above is called a *secondary current;* the voltameter acts as a *secondary battery.* The secondary current thus obtained lasts but a short time, and its intensity is seen to diminish rapidly from the moment it begins to circulate in the galvanometer and is soon reduced to nothing.

We will again have occasion to speak of secondary batteries, of which we have just given an example, and which have lately undergone vast improvements.

POLARIZATION OF A VOLTAIC CELL.

If the current furnished by a voltaic cell (one of Wollaston's, for instance) with well-amalgamated zinc be examined by means of a galvanometer, the intensity is seen to diminish from the moment the circuit is closed.

This diminution is very rapid if the circuit has but very little resistance; it is, on the other hand, very slow if the circuit offers great resistance, as in a long line of telegraph.

If, after having allowed the current to flow for five minutes, for instance, the circuit be left open for five minutes, it will be seen when again closed that the current has nearly assumed its first intensity. It can be said then, the battery when not at work regains its initial power.

It may be understood from these observations how it has been possible to use the sand-battery for a number of years in the telegraph service; the telegraph lines offering indeed great resistances, but only needing intermittent currents.

By closely examining that which takes place while the circuit is closed, different circumstances of the phenomenon will be seen, which will throw a great deal of light upon the causes to which it must be attributed.

At first bubbles of hydrogen are seen to form themselves upon the copper electrode, as we have already stated; this will lead to the belief that imperceptible bubbles form themselves upon the entire surface in such a way as to interpose, more or less completely, between the electrode and the liquid, a gaseous layer. Thus apparently the principal cause of the diminution in the intensity of the current should be sought at the surface of the copper electrode.

Several simple experiments will confirm this.

If, after a marked diminution in the deflection of the galvanometric needle, the electrodes be shaken without lifting them out of the liquid, the current is seen to partly recover the force it had lost.

The same thing is observed if the liquid alone be shaken without moving the electrodes, and consequently without changing the extent of the immersed surface.

The moving of the copper electrode alone will show, as a result, the recovery of the lost force.

By rubbing the copper, without taking it out of the liquid, with a small brush, the same result is noticed.

In these three experiments the disappearance of bubbles of hydrogen from the surface of the conducting electrode is accompanied by a renewal of the intensity of the current.

If, on the other hand, the zinc electrode alone be agitated, no perceptible modification in the decrease of the current takes place.

Henceforth there can be no doubts as to the importance of the phenomenon which takes place at the surface of the copper electrode. The diminution of intensity that we have observed may be attributed to two causes: either to the increase in the internal resistance of the battery, or to the decrease in the electro-motive force. In fact, the two causes are present at the same time. That the resistance increases cannot be doubted, since the active surface of the copper electrode is diminished; but a simple and direct demonstration of this does not seem easy to obtain.

That the electro-motive force is diminished is extremely easy to demonstrate. For this experiment we employ the method of opposition which we have already described, and which is as convenient for the comparison of electro-motive forces as are scales for the comparison of weights.

The instant the electrodes are immersed in the liquid and the battery begins to work, it attains its maximum intensity.

Let us now take two identical battery cells and close the circuit of one of them for five minutes, leaving the other inactive. At the expiration of five minutes, place the one that has been working in opposition to the fresh one, and a galvanometer interposed in the circuit will show the superiority of the electro-motive force of the fresh cell.

If now these two cells be made to act separately, each upon itself—that is, without the insertion of any resistance during five minutes—it will be found at the end of that time, by placing them in opposition, that the second one still has a greater electro-motive force than the first one.

The experiments could be varied, and it could be ascertained, for instance, how long the decrease continues in a cell of a certain size and form and under well-known circumstances.

It can be easily shown that the electro-motive force of a voltaic cell can, by constant action, be reduced one half. For this it is only necessary to cause two cells to work a considerable length of time; when they are supposed to be exhausted as much as they can be, join them in intensity and place this battery of two cells in opposition to an entirely new cell; the galvanometer will still mark the superiority of the latter, and the necessary conclusion is that the electro-motive force of each one of the two exhausted cells has been reduced to less than half of that of the new cell.

It is admitted that the diminution in the electro-motive force of batteries is due to the production of an electro-motive force (upon the surface of the negative electrode) contrary to that of the principal current.

This view is founded upon that which we have said of the electro-motive force found in a voltameter, from whose electrodes gases are given off.

It may be shown by a direct experiment that the conducting electrode C of a weakened battery has acquired peculiar properties. It is only necessary to immerse in the liquid a second plate of copper, C', and to connect the two with a galvanometer. The passage of a current is thus made manifest, and its direction shows that the copper plate C acts as the soluble electrode, or electro-positive, when compared with the other, C', which assumes the part of a conducting electrode, or electro-negative. This current commences to decrease from the moment it is established, and soon becomes imperceptible. Thus the electrode C, which was electro-negative in the voltaic cell before and during its weakening, is electro-positive in the test cell of two copper electrodes. Finally, if after the above experiment the voltaic cell be re-established, it assumes its original intensity, at least for a moment, and then begins to weaken again, as in the first instance.

It is then that the *conducting electrode* is said to be in a state of *polarization*.

Such is the phenomenon of the polarization of the negative electrode of batteries, a knowledge of which is so important.

It will be seen, in the following pages of this work, that the less the polarization, the better the batteries. The most important improvements in batteries are those which have had in view the diminution or suppression of polarization. In other words, the principal aim and effort of inventors worthy of that name has been to *depolarize the electrode*.

It has been established that *polarization* remains the same when the size of the cell and the intensity of the current are in proportion to each other. It is here necessary to define polarization: *it is the difference between the electro-motive forces in a polarized battery and a depolarized battery*.

It can be understood indeed that the quantity of hydrogen given off upon the negative electrode is in proportion to the intensity of the current; and that if this quantity distributes itself upon the surface of an electrode also proportional, the thickness of the deposit will be the same, and consequently its intrinsic action will not have changed. The practical conclusion of this law is that polarization will be less in a battery having large electrodes than in a smaller one, although the total resistance be the same.

POLARIZATION IN A BATTERY OF SEVERAL ELEMENTS.

Thus far, each time that we have spoken of the polarization of the negative or conducting electrode of cells, we have implicitly supposed the cell to be alone, and that the current which produced the polarization was the current of the cell itself. In ordinary practice it is not thus; several elements are generally joined in intensity, and the current which flows in each one is furnished by the entire battery.

Let us place 10 cells, each having 10 units of resistance, in a circuit of 100 units (total resistance 200 units); it is clear that the current will be more intense than if 9 of the 10 cells were taken away; consequently the current which produces the polarization in each cell will be more energetic than if there were only one cell.

The result is that the weakening due to polarization is more marked in cells which are joined in intensity than when they are separate.

In other words, when a current, passing through a cell, is more energetic than the current which the cell itself produces, the weakening of the current takes place under the following circumstances:

At first hydrogen is given off upon the copper, and

produces that which we have termed polarization of the cell.

But afterwards, when the greater part of the acid is converted into sulphate of zinc, the sulphate itself becomes electrolyzed and reduced zinc deposits itself upon the copper. If at last this deposit covers the entire surface of the copper, it can be easily seen that the two electrodes will become identical, and consequently it is no longer a battery cell.

We shall show instances where some of the cells of a battery not only cease to produce a current in the right direction, but actually produce a reverse current.

CHAPTER IV.

SULPHURIC-ACID BATTERIES.

At the point which we have now reached we are able to compare different batteries and to undertake their study.

Up to this time we have only shown Volta's battery and the modifications in its arrangement. We will now examine batteries which are analogous, but which differ more and more from the first model.

This study will show how Volta, in spite of his imperfect means, happily chose the elements which have been used ever since; it will be seen how advantageous and how imperative the use of zinc is.

We will first study those batteries in which the liquid is dilute sulphuric acid, but in which the electrodes differ from those in the voltaic battery.

BATTERIES WITH CARBON ELECTRODES.

A battery differing from Volta's only in the substitution of carbon electrodes for those of copper is very often employed; it was invented by Mr. Walker in 1849.

In these cells the negative electrodes are made of gas carbon, which forms a shell upon the heated retorts in the preparation of gas. This substance has a very good conducting power, and it is very porous. On account of this porosity the electrode presents a considerable surface, and is very slowly polarized.

We have already explained, in speaking of Wollaston's battery, why it was advantageous to give the largest surface possible to the conducting electrode from which hydrogen is given off. The method that we have given to show the progress of polarization in a battery cell proves the superiority of a battery with carbon electrodes over that of Volta of equal dimensions.

The zinc may be placed between two plates of carbon, or better still in the centre of a hollow cylinder of carbon, always having in view the increase of the surface to be polarized and the checking of the polarization.

MANUFACTURE OF CARBON ELECTRODES.

When carbon electrodes have simple geometrical forms, or when they are simple plates more or less thick and wide, they may easily be cut from the residue in gas-retorts, and that is what is generally done.

But if they are cylindrical, and especially hollow cylinders like that shown in Fig. 18, the above process cannot be applied. The electrodes must be produced artificially in moulds, by pressing powdered carbon with proper cements.

Bunsen suggests the following process to make carbon:

A mixture of one part by weight of coal and two of coke is made (both having been reduced to an impalpable powder), which, placed in a sheet-iron mould, is heated to clear red until all gases have been given off. The carbon is then dipped in molasses and left to calcinate, protected from the air.

John T. Sprague, of Birmingham, recommends the following process:

"Plates or blocks may be built up from powdered

graphite mixed up with coal-tar or strong rice-paste into a stiff dough, which should be dried, heated, then packed in powdered carbon in a closed vessel and heated to clear red for some time. When cool they should be soaked in strong syrup of sugar, or treacle, again dried and treated as before; this process must be repeated until the carbon is perfectly dense and strong."

Fig. 18.

USE OF CARBON ELECTRODES.

The chief difficulty with carbon is in making the connection. The contact between the carbon and the metallic *rheophore*, by which it is connected with the adjoining cell or with the circuit, must be perfect.

This is commonly done by fixing a clamp on it to which the *rheophores* are attached.

A better plan is to deposit copper on the upper part and then solder the connection to it, as this gives continuous circuit. There is one drawback, however: the acid is soaked by capillary action into the pores of the substance, reaches the surface of the carbon and the inner surface of the copper, which it attacks, thus destroying the connection. It is easy to avoid this action by immersing the upper part of the carbon in melted paraffin. The pores of the immersed part are thus filled by the paraffin, which, when left to cool, becomes solid. All capillary action through the upper part of the carbon is thus prevented.

The top of the carbon may also be immersed in melted zinc. But by capillarity the liquid can ascend and attack the zinc as it did the copper. The sulphate of zinc would present the same difficulties as the sulphate of copper, and it is also desirable in this case to dip the upper part of the carbon in paraffin. The experiment shows that paraffin does not affect the conductivity of the carbon, and that the resistance of the battery is not increased by this addition.

Lead may also be deposited upon the top of the carbon, but here the paraffin is indispensable, because the formation of sulphate of lead is enough to diminish considerably the intensity by introducing in the current a matter almost without conductivity and nearly insoluble.

In Switzerland the battery which we have above described is extensively used, especially in telegraph offices. The zinc should be well amalgamated before being placed in the centre of the carbon cylinder (Fig. 18), in order to diminish local actions while the battery is at rest; owing to this precaution the battery may be used a long time without any care being bestowed upon it.

SULPHURIC-ACID BATTERIES.

We will see farther on how this battery has been improved upon by substituting a solution of sea-salt for the dilute sulphuric acid.

ZINC-IRON BATTERY.

One of the first ideas, and the most natural, is to use iron on account of its cheapness. Iron may indeed be substituted for the copper, but a battery thus arranged is very inferior to that of Volta. The substitution of iron for copper causes a notable diminution in the electromotive force. It is important to note, however, that the copper may be effectively replaced by iron; that it is still the zinc which is attacked; and that the iron is preserved from the action of the sulphuric acid while the circuit is closed.

IRON-COPPER BATTERY.

In Volta's battery it is the zinc which is continuously dissolved; it is therefore logical to search for something which may replace the zinc and which at the same time is less costly—iron, for instance. This substitution of iron for zinc would be more advantageous than the substitution of iron for copper; but this battery (iron, copper, and sulphuric acid) is still inferior to the preceding one.

OTHER COMBINATIONS.

If the question of economy be put aside, many other combinations might be usefully employed; but, as we have said, the use of zinc is necessary, as no other metal practically acceptable can be advantageously substituted.

Even aluminium is less liable to be attacked, or, as it is said, is less *electro-positive* than the zinc. Only calcium, sodium, potassium, and analogous metals are more electro-positive. It is needless to say, however, that they cannot be used in batteries destined for practical purposes.

For the negative or insoluble electrode there is, on the other hand, great choice : lead, silver, and platinum can be and are often employed. The electro-motive force of a zinc-platinum battery is rather superior to that of Volta's (zinc-copper), and is about equal to the zinc-carbon battery.

Following is a list of metals so arranged that if any two be taken to form the electrodes of a dilute sulphuric-acid battery, the one nearest the end of the list will be the positive electrode, or the negative pole of the cell thus arranged :

1. Silver.
2. Copper.
3. Antimony.
4. Bismuth.
5. Nickel.
6. Iron.
7. Lead.
8. Tin.
9. Cadmium.
10. Zinc.

Too great an importance must not be attached to this list, for the order of the metals would be different if the liquid were other than dilute sulphuric acid.

SMEE'S CELL.

Many ways have been devised for reducing the polarization of the negative electrode of the batteries which we have described. In 1840 Smee indicated a very ingenious way, which consists in using electrodes of platinum, upon whose surface he deposited, by means of electricity, platinum as a fine black powder. These electrodes of

SULPHURIC-ACID BATTERIES.

platinized platinum tend to diminish considerably polarization. The simple reason of this is that the bubbles of hydrogen free themselves much more easily than from the polished surface of a metal.

For reasons of economy Smee placed the platinum plate between the two plates of zinc. It is in form a reversed Wollaston battery.

Smee's battery is charged with a solution containing one part of acid to seven parts of water. Its work is much greater than could be expected from a single-liquid battery.

Again, for economy, Smee replaced the platinized platinum by platinized silver. The following composition

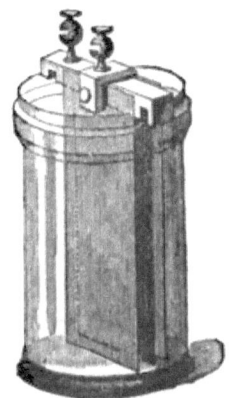

FIG. 19.

had even been used, which produces a much cheaper cell:

Upon a plate of copper is deposited a grainy layer of copper, then a layer of silver, and finally a layer of platinum dust. The rough surface thus given to the silver facilitates the deposit of platinum, which is very difficult upon polished silver.

One of Smee's batteries would give very unsatisfactory results if the zinc were not amalgamated; it is a precaution that should not be neglected.

This battery is extensively used in England and the United States with many modifications, one of which is presented by Fig. 19.

WALKER'S PLATINIZED CARBON BATTERY.

We have stated above how, since 1849, Walker had used batteries with electrodes of carbon cut from the gas-retorts. In 1857 he resolved to platinize his carbons, and the battery thus constructed has been used by the South-Eastern Railway in England with great success; nine thousand of these cells were in service in March, 1875.

These cells are contained in an earthen jar, and the lower extremity of the zinc is immersed in a gutta-percha saucer filled with mercury; the zinc is well amalgamated, which reduces to its minimum the local action or lost chemical work; the top part of the carbon is copper-plated and tinned.

The usual size of these cells is 4 inches by 2 inches; their price (with the mercury and sulphuric acid), 42 cents.

The cost of keeping them in order is calculated at 25 cents annually.

This battery may be left twelve, fifteen, and sometimes seventeen months without needing any care whatever.

It is a simple modification of Smee's battery, and with a liquid of one part of acid to eight parts of water there is an electro-motive force equal to that of Smee's (measurements made before any polarization). Polarization may reduce the electro-motive force one half. It will

SULPHURIC-ACID BATTERIES.

be seen from the tables at the end of this work that this force is equal to that which is taken as the unit; namely, that of Daniell's battery.

The internal resistance of Walker's battery is about 1 ohm, or 1 unit. It is certainly a very small resistance for a telegraph battery, a quality which we must point out.

TYER'S BATTERY.

Tyer combined a modification of Smee's batteries, for the service of electric railroad signals, which presents many advantages.

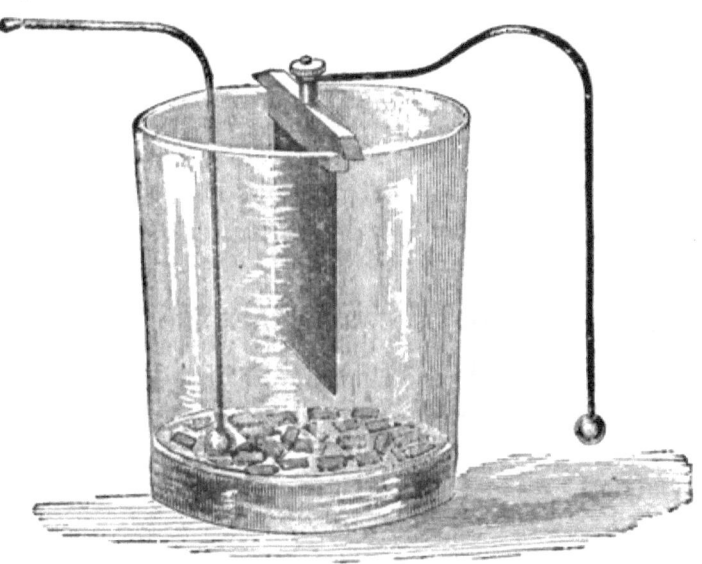

Fig. 20.

In the bottom of the jar (Fig. 20) are placed a sufficient quantity of mercury and pieces of zinc; this constitutes the generating electrode.

A plate of platinized silver is held vertically in the jar

by means of a cross-piece of lead which rests on the rim of the jar, thus giving a good height and a certain firmness to the conducting electrode. The top of the cross-piece of lead is furnished with a terminal, to which is fastened a copper wire covered with gutta-percha; at the end of this wire is a ball of zinc which is wholly immersed in the mercury of the adjoining cell.

The liquid is sulphuric acid diluted with twenty times its volume of water.

This battery has the advantage of consuming fragments of zinc and using them to their last particle. Those pieces which are wasted in the manufacture of other batteries can here be put to use. In this respect Tyer's is the best arrangement yet produced.

The maintenance of this battery is reduced to a minimum, for in a well-closed box it can remain two or three years without examination. Great care should be taken, however, in the charging and cleaning of the battery, in order to avoid any loss of mercury.

BARON EBNER'S BATTERY.

To the Austrian general, Baron Ebner, is due the following arrangement of Smee's battery. The negative electrode is of platinized lead; the generating electrode is, as in the preceding arrangement, composed of fragments of zinc in some mercury, which keeps them well amalgamated.

A very large battery of this kind was used at the Paris Exposition of 1867 to run electric clocks; which proved that the polarization was but slightly felt, as electric clocks do not work well unless a very constant current is provided.

The electro-motive force of this battery is only about

half that of Daniell's battery, of which we will speak farther on, and which is generally taken as a term of comparison. Its maintenance is very economical, for the same reasons given in the description of Tyer's battery.

BATTERIES ANALOGOUS TO THAT OF SMEE.

Following Smee's example, Poggendorff deposited pulverized copper upon a copper electrode and thus obtained a battery of Volta, or of Wollaston, notably improved, inasmuch as the polarization takes place less rapidly and with less intensity.

Drivet, an Italian officer, carried out an analogous idea. He deposited upon the copper electrode of a voltaic cell a very thick layer ($\frac{1}{8}$ of an inch) of spongy copper. The porosity of this metal gives it some of the qualities of carbon electrodes. The analogy with Smee's battery is more apparent than real, for the very thin layer of pulverized platinum does not present the increase of surface which is the advantage in the use of carbon electrodes.

We have ourselves tried a battery in which the negative electrode is a plate of lead, upon whose surface a layer of spongy lead $\frac{1}{25}$ of an inch thick is deposited.

REMARKS UPON POLARIZATION IN THE PRECEDING BATTERIES.

We have seen in all batteries described thus far that polarization was the result of the freeing of gaseous bubbles of hydrogen from the negative electrode.

We have indicated several means, devised by different physicists, to diminish this effect, which is done either by increasing the surface of the electrode to be polarized (Wollaston's battery, carbon-electrode battery, and Drivet's battery) or by modifying this surface in such a way as to facilitate the freeing of the gas (Smee's and similar batteries). The action of a battery is already vastly improved by giving a rough surface to the polarized electrode, instead of leaving it polished.

We have also shown how the air acts favorably upon batteries, either by diminishing polarization while they are at work or by producing depolarization when the current has ceased to flow. Depolarization would undoubtedly take place in the absence of the oxygen of the air, by the freeing of gas or by its dissolution in the liquid; but the oxyen renders depolarization much more rapid, especially in the case of carbon electrodes, by combining with the hydrogen to form water.

We will indicate, as we proceed, much more effectual means for diminishing or suppressing polarization, which consist in the use of substances placed near the negative or conducting electrode, and by which the hydrogen is chemically absorbed. These are the only contrivances by which constant batteries can be produced; that is, batteries whose electro-motive force is constant.

We will describe in detail these constant batteries, which present a satisfactory solution of the problem of obtaining a continuous and regular electric current. They have taken the place of simple and inconstant batteries in all applications, and their study will be the crowning of the present work.

But in order to proceed from the simple to the complex we ought to describe several other inconstant batteries, only a few of which have any practical interest. Their study is, however, necessary in order to understand the many varieties already tried and those which might be tried.

CHAPTER V.

ACID BATTERIES ANALOGOUS TO THAT OF VOLTA.

Thus far we have considered a series of batteries differing very little from each other, all being composed of two different electrodes immersed in a single liquid, dilute sulphuric acid.

It is easily understood that by replacing the sulphuric acid by other acids, new batteries analogous to the first ones may be obtained.

HYDROCHLORIC-ACID BATTERIES.

The cheapness of hydrochloric acid caused many persons to use it; but none of the batteries thus constructed obtained any continued application, because hydrochloric acid, being gaseous and only soluble in water, escapes into the surrounding air, so that after a short time it is impossible to remain in the room where it is placed.

Besides, the hydrochloric acid liberates itself rapidly from the water in which it is dissolved, at least a good part of it, the liquid becoming immediately impoverished, and a new cause of the weakening of the current is added to those which we have already pointed out.

NITRIC-ACID BATTERIES.

Nitric-acid batteries could be very easily made, but they would have the same inconveniences as those with hydrochloric acid, and would not present the same economical advantage. It will be seen, however, that in certain less rudimentary combinations nitric acid is put into use.

VARIOUS ACID BATTERIES.

All acids employed by chemists may be used in the composition of batteries, provided they be liquid or soluble in water and conductors of electricity.

Acetic acid, found in all households, may be used in the absence of others. It has indeed been used by Pulvermacher in his electro-medical battery. The electrodes were zinc and copper wires wound upon small pieces of wood. They were connected with each other, the positive pole of each with the negative pole of the following one, and dipped in diluted vinegar. Twenty years ago this apparatus had great success, but to-day it is replaced by others more perfect.

In all these voltaic combinations the chemical action is the same as in Volta's battery. The zinc becomes oxydized at the expense of the water, and the oxide of zinc combines with the acid, forming a nitrate, an acetate of zinc, etc. The hydrogen of the water is given off upon the negative or conducting electrode.

It can be seen without going any farther how many different batteries may be conceived by simply varying the nature of the electrodes and the liquid. But many of these combinations are far from possessing any interest, and our remark is only designed to call the attention of the reader to the number of solutions of the problem of constructing batteries.

CHAPTER VI.

BATTERIES WITHOUT ACIDS.

IN addition to the acids there are numbers of liquids or solutions which may be used in batteries, a few of which are interesting.

SEA-SALT BATTERIES.

On account of the facility in obtaining chloride of sodium, or sea-salt, or common salt, it is often made use of in batteries. The battery, whose electrodes are carbon and zinc, is almost exclusively used in Switzerland for telegraph purposes, with dilute sulphuric acid, or more frequently with salted water.

There are several dimensions of these. The smallest has flat electrodes $2\frac{3}{4}$ inches long; the next size has electrodes 4 inches long and $1\frac{1}{2}$ inches wide. Both sizes have but a single piece of carbon in each cell. The first can work one month, the second three months, without care.

These cells certainly cost very little, and there is scarcely any consumption of the zinc while the circuit is open, although the zinc is not amalgamated, which is a very satisfactory condition. The electrodes may be lifted out during the suspension of work, and this is facilitated by the electrodes being attached to a bar of wood. By lifting this bar the electrodes of ten cells may be raised at one time.

Another model of this same battery is shown in Fig. 18.

The carbon has the form of a hollow cylinder, in the centre of which is a plate of zinc, not amalgamated, as we have already stated; these two electrodes are fastened to a strip of wood which rests upon the rim of the jar containing salted water. The comparatively large surface of the carbon is a very favorable condition (for single-liquid batteries), as we explained when speaking of Wollaston's battery. In the model employed on Swiss lines the carbon is $5\frac{1}{2}$ inches high, and has an exterior diameter of $3\frac{1}{2}$ inches. These batteries can do service from nine to twelve months without requiring attention.

The friend to whom we are indebted for the preceding information uses salt-water batteries for domestic bells. He employs cells which have a height of 14 inches. Some batteries of this kind have been known to work from six to eight years without any care whatever. There is one which worked ten years; the zinc had of course disappeared.

Concerning the weakening of this battery, it has been found that it may be exhausted by causing a constant current of a short circuit to pass for ten or twelve hours, and that it only needs two or three hours of rest to regain its lost energy. In other words, depolarization takes place very rapidly.

The sea-salt battery is not only used for the telegraph and electric bells, but for electric clocks.

DUCHEMIN'S ELECTRIC BUOY.

Duchemin placed elements of the preceding form directly in the sea by attaching them to some floating body. The constant agitation of water caused undoubtedly an almost complete depolarization. When several cells are employed, however, they are in the same liquid; there is therefore a small loss of electricity; it cannot be of much consequence, because of the form of the carbon which surrounds the zinc. A perfect insulation of the wires connecting the cells is of great importance.

The main object of these batteries was the preservation of sheets of iron used in the construction of vessels, barges, buoys, etc. etc. It appears that the hull of a vessel undergoes a relatively less change during navigation than when at anchor or in port; it is in this case that the use of Duchemin's buoy is practicable. These buoys are used as follows:

Seven cells about 4 inches in diameter, for instance, are joined in intensity. The positive pole of this battery is put into communication with the sheets of iron to be preserved; the negative pole (that is, the zinc of the last cell) is in the sea, as indeed are the others. Under these circumstances it has been proved, by experiments made at Cherbourg by officers of the French navy appointed purposely by the Minister of Marine, that a surface of iron eighteen times larger than that of the zinc which forms the soluble electrodes of the battery may be preserved from rust.

It appears that the simple addition of a sheet of zinc is not sufficient to preserve the hull of an iron vessel from rust, but it can be done by means of one or several electric buoys; that is the result, at least, of a prolonged experiment on a small iron boat.

These interesting experiments were unhappily discontinued during the war of 1870–71, and have not been resumed.

Before leaving the subject, we will say that there is a possible superiority of sea-water over common salted water; for sea-water does not contain chloride of sodium alone. We have unfortunately no positive information upon this point.

The salted-water or sea-water battery, although inferior to many others (especially to the sal-ammoniac battery, of which we will speak later), may be recommended, above all, in places near the sea, where the expense is comparatively small.

ZINC-COPPER-SEA-WATER BATTERY.

At the beginning of the present century the illustrious Sir Humphry Davy proposed to protect the copper hull of vessels by means of a sheet of zinc (or, indeed, of cast-iron) put into communication with the lining and immersed with it in the sea. A cell was thus constructed in which the zinc (or iron), by being attacked, protected the copper.

The zinc had, of course, to be replaced at the end of a certain time; but an extent of active zinc surface one hundred and fifty times larger than that of the copper was sufficient to protect the latter. This ingenious idea had to be abandoned in the practice for the following reason:

The zinc-copper cell of which we have spoken gives off, indeed, hydrogen upon the surface of the copper; but at the same time it decomposes certain salts contained in sea-water, and the bases (earthy oxides—magnesia and lime) deposit themselves upon the copper. To this crust sea-grasses and shell-fish attach themselves and slacken to a great extent the speed of the vessel.

In the absence of the zinc the copper is slightly attacked by the sea-water, but the surface remains apparently clean. In the long-run the copper is used up; but of two evils one must choose the less, and prefer to lose a little more on the resale of old linings than to increase the duration of voyages.

ZINC-IRON-SEA-WATER BATTERY.

Within the last twenty or thirty years copper-lined vessels have gradually been abandoned and a great number of iron ships have been constructed. Davy's idea is in this instance applicable. We do not know if many experiments have been made or not; but the result of one experiment upon a French frigate showed that sheets of metal one metre square lost the following weights after remaining in sea-water one month:

	Grammes.		Grammes.
Steel	28.10	Zinc	5.60
Iron	27.30	Galvanized iron	1.80
Copper	3.80	Tin	1.50
Lead	only traces.		

These figures go to prove that iron is of all metals the most attacked by sea-water, and is therefore badly chosen, as far as preservation is concerned, for the construction of ships, buoys, etc.

From a theoretical point of view there would be a great advantage in coppering or tinning iron. If the iron were thoroughly covered with a thin layer of copper or tin it would no longer be in contact with the water, and would consequently not be attacked. But a small accident, such as the scraping of the ship on a sand-bar, for instance, might be enough to chip off a little piece of copper or tin, when the exposed iron would immediately be attacked.

Thus would be established an iron-copper or iron-tin cell which would excite the action of the sea-water upon the iron. It might indeed go so far as to make a hole in the iron. It will be seen that the cell thus formed would possess peculiar conditions of activity, as the negative electrode is enormous when compared with the soluble electrode; and besides, the constant agitation of the water would tend to suppress all polarization.

In the experiments above referred to the surface of the sheets of metal were, of course, well cleaned before each experiment.

There is no doubt as to the zinc being the electro-positive element of the zinc-sea-water-iron battery, and consequently that the iron is electro-chemically protected by the zinc.

It is possible that the feeble electro-motive force of this cell may be insufficient for a thorough protection. There may also be some accessory action which might make the action of the cell worse than simply ineffectual, as in the case of the copper linings.

This is, we think, a desirable question to elucidate.

ACCIDENTAL REVERSING OF THE CURRENT.

We have already shown how a voltaic cell may be rendered ineffective by electrolysis of the salt of zinc and the deposit of zinc upon the conducting electrode.

We have said that in certain cases the current could be reversed; this phenomenon was observed under the following circumstances:

Certain zinc-salt-water-carbon batteries that had been working two years were, by accident, short-circuited; polarization was brought to its maximum, since there was no resistance in the external circuit, and consequently the intensity was the greatest it could be. Soon after the battery could supply almost no current whatever; and by close examination it was found that in one out of every four or five cells the poles were reversed; that is, the zinc had become the positive pole, and the carbon the negative pole. There was present a polarization similar to that which would have taken place in a voltameter, or in a secondary battery placed in the circuit. This secondary current neutralized, in a great measure, that of the other cells of the battery.

In this singular instance the polarization of certain elements had become stronger than the element itself.

It is easily understood that, if there had only been one cell in the circuit, this reversing of the poles would never have taken place; for the current resulting from polarization is necessarily inferior in tension or electro-motive force to the polarizing current.

One more remark before leaving this experiment. If the battery cells joined in intensity were and would remain identical, the above phenomenon would not take place.

If twenty cells were joined in intensity and in short circuit (that is, without any exterior resistance), the intensity is exactly the same as if there were but one cell in short circuit; for in the first instance the electro-motive force is twenty times greater and the resistance of the circuit twenty times less than in the second instance, which establishes an exact compensation.

If there be only one cell in the circuit, it cannot but

weaken, and no reversing of the poles can take place. Therefore in a battery whose cells are identical there can be no reversing. For the occurrence of this phenomenon the cells must necessarily be dissimilar, which is nearly always the case. Some are polarized from the beginning much more rapidly than others; from that time they are no longer identical cells, and the poles of the weaker ones, which are the most polarized, may be reversed.

If some of these cells should accidentally be closed while the others are open, they become rapidly polarized; the cause of this may be the formation of climbing salt or other causes. This remark shows the advantage of the careful cleaning of batteries, in order that they may work regularly and for a long time.

We will return to this subject when speaking of the sulphate-of-mercury battery.

CHEMICAL ACTION IN SEA-SALT BATTERIES.

No one, as far as we know, has analyzed the products formed in this battery; it must be a very complex composition, a mixture of chloride of sodium and oxide of zinc, or of soda and zinc chloride. There can only be conjectures upon this subject, as no analysis has been made. The only known fact is that hydrogen frees itself from the carbon.

These analyses are probably very difficult, and the combinations formed in batteries are in general very complicated; the slowness of the actions favors the production of bodies more complicated than those of which mineral chemistry generally treats.

Batteries may some day challenge the particular attention of chemists, who will find, without doubt, that they are as good as retorts and other apparatus used in laboratories for the formation of composed bodies, of which a considerable number have not yet been studied. We are sure that chemistry will lose nothing, and it is certain that the science of electricity will be greatly benefited by this study.

The difficulty of the chemical problem presented by the sea-salt battery, and indeed by nearly all batteries, is increased by the fact that the nature of the compositions formed is different when the current is closed, and when it is open. It is certain that if there were a change in electric conditions, the action of affinities would also change.

Our attention will again be called to this subject, when we will give certain reasons supporting the above suggestion.

MARINE BATTERIES.

An old experiment shows that if a plate of zinc and a plate of copper be immersed in the sea at a considerable distance from each other and attached to a single conducting wire, there will be produced in this wire a current of considerable intensity. Whichever way it may be looked at, the internal resistance of this battery is very feeble; either by considering the sea as the jar containing the liquid and the two electrodes; or, adopting recent views, by admitting that the electricity is lost in the earth (the common reservoir) at those two points where the line touches it. This combination is not susceptible of practical application, as it only furnishes one cell and not a multiple battery; but from a theoretical point of view it deserves notice.

SAL-AMMONIAC BATTERIES.

By substituting a solution of chloride of ammonium or sal ammoniac for the liquids previously mentioned, a new series of batteries, analogous to those already enumerated, may be realized. We will call the reader's attention to but two of them, which possess particular interest.

BAGRATION BATTERY.

The electrodes of this battery are the zinc and the copper; they are immersed in a jar filled with earth sprinkled with sal ammoniac. "It produces a wonderfully constant current which is the result either of the reduction of the hydrogen upon the copper by the composition formed there by the sal ammoniac, or of the absorption of the hydrogen by the earth itself, which indeed acts as a diaphragm. It is best not to put the two plates of the cell too near to each other, and to immerse the plate of copper, before putting it in the earth, in a solution of sal ammoniac, leaving it to dry until a greenish layer is formed upon its surface." *

In spite of these precautions, this battery has been put aside; but it is possible that it may again be taken up.

CARBON-ELECTRODE BATTERY.

This battery differs from the preceding one in the substitution of carbon for copper. We have already explained the advantages of carbon.

To give to these batteries their maximun force, and to

---

* De La Rive, Traité d'Électricité.

BATTERIES WITHOUT ACIDS. 73

render polarization as slow as possible, they are arranged as follows:

The carbon electrode is placed in a porous porcelain jar, which is then filled up with small pieces of carbon; a considerable extent of surface is thus given to the negative electrode. This porous jar is then placed in a glass or stoneware jar which contains the solution of sal ammoniac. The zinc is immersed in the liquid; it has the form of a hollow cylinder, and a thickness of $\frac{1}{25}$ of an inch is sufficient, as there is but little waste.

This battery presents an important advantage, only found thus far in the Bagration battery and in the sea-salt battery.

All batteries, indeed, in which the positive or soluble electrode is zinc in sal ammoniac have the same advantages.

As long as the circuit is open there is no chemical work going on; the action of the sal ammoniac upon the zinc does not commence until the circuit is closed, and ceases immediately upon the reopening of the circuit. In order to make this important point perfectly clear, the following experiment must be made: Place an ordinary piece of zinc in a solution of sal ammoniac and leave it there for some time, several weeks for instance, when it will be seen that the zinc is not attacked in the slightest degree. If now a fragment of metal, iron, or copper, or even a piece of carbon, be added in the jar, it is soon seen that the zinc is attacked and a white salt is formed.

It is thus seen that for the attack of sal ammoniac upon the zinc the formation of a cell is necessary; if the zinc does not touch the metal added, no attack takes place.

We will not insist upon the many consequences of this

simple experiment, but the plain result is, that, in batteries formed with zinc immersed in a solution of sal ammoniac, there is no chemical action except when the battery is doing useful work.

From a practical point of view this advantage of sal-ammoniac batteries is capital, for in most applications batteries only work at intervals. In the principal telegraph offices, where the greatest number of telegrams are sent and received, only intermittent currents are used, but in such quantities that an almost constant demand for electric current is imposed upon the battery; but in branch offices there are generally long intervals between telegrams, and there is most frequently no service at all during the night.

In the application of domestic bells, the battery should always be ready for work night and day; the current is consequently used, on an average, but a few minutes in the twenty-four hours. In applications of this kind it is seen that the period during which the batteries remain inactive is a hundred times, nay, two hundred times, longer than that during which they work, and that the economy should therefore be made during the time in which no current is required.

We will add that, such as it is, it could be very well used for electric bells, and could, if necessary, serve in a branch telegraph office. The battery in question has indeed but one fault: it becomes polarized when furnishing a current. However, for such an intermittent service this inconvenience disappears; for during its short period of work, polarization is barely perceptible, and it has sufficient time to disappear completely during the long intervals of rest. To M. Leclanché is due the discovery of the advantages presented by the sal-ammoniac battery. He first established the fact that a battery could be produced in which the waste did not exceed, in proportion, the electricity supplied.

Another advantage of this battery is that if, at a certain time, it is seen to weaken and there be no sal ammoniac at hand, it can be charged for the time being with common salt. But this means should only be resorted to in an emergency, as the current obtained with common salt is less intense than that furnished by sal ammoniac.

ACTION OF AIR UPON THE PRECEDING BATTERY.

From various experiments made with carbon-electrode and sal-ammoniac batteries the following conclusions may be drawn:

1. The surface of the carbon should be as large as possible compared to that of the zinc; and by increasing the mass of carbon according to a given quantity of zinc, polarization may be suppressed.

2. A part of the carbon should be exposed to the air; for it has been proved that when the carbon is totally immersed the intensity is diminished, but recovers as soon as some of the liquid has been taken out. This is what a French physicist calls *letting the carbon breathe.* The use of porous jars which overreach the top of the glass jar is very important.

3. Preference should be given to gas-retort carbon, on account of its porosity, and, as we said in speaking of the chemical action in Volta's battery, it must be used in fragments large enough to permit the access of air. The powdered cake, formerly used, should be discarded.

These conclusions will be readily admitted by the

reader, who can understand that the presence of oxygen in the pores of the carbon contributes to the depolarization of the battery. It is possible that the particular faculty of the carbon for absorbing gases in large quantities here plays some part, and that the properties of the gases thus condensed in the pores of the carbon may be different from what they are under ordinary circumstances.

CHEMICAL ACTION IN SAL-AMMONIAC BATTERIES.

A French chemist, in analyzing crystals formed in sal-ammoniac batteries, found this formula for them:

$$3ZnCl, \ 4NH_3, \ 4HO.$$

The gases given off from the element were found to be:

$\frac{1}{2}$ volume of hydrogen.
$\frac{1}{8}$ " nitrogen and carbonic acid.
$\frac{1}{8}$ " heavy carburetted hydrogen.

These results only confirm that which we have said above of the complicated composition of bodies formed in batteries.

OTHER BATTERIES.

ZINC-IRON-WATER BATTERY.

We have already spoken, several times, of batteries in which the electrodes were zinc and iron; and we have seen that the zinc was always the generating electrode, and the iron the conducting electrode.

Every one knows that to protect iron from rust it is covered with zinc, and is then generally known under the name of *galvanized iron*.

If the galvanized iron be exposed to rain or humidity, the uncovered parts of the iron constitute a cell with the zinc, and the iron is protected by the zinc. This electro-chemical protection increases considerably the importance of the process of galvanizing iron.

It must be said that the oxide of zinc produced by the exposure of the zinc to the air is insoluble in water, and forms a protecting layer which hinders further oxidation; that is the principal reason for the use of zinc in out-door works alone or with iron.

IRON-TIN BATTERY.

It is interesting to examine, from the same stand-point, tinned iron. Tin is liable to but very little alteration in water or when exposed to damp air. For many years it has been the practice to tin iron, by which its durability is greatly increased.

It is important to note that if the iron be exposed at any point it is promptly attacked, because in the voltaic cell formed with water, or simply damp air, the iron is the generating electrode. Under these circumstances the rust is seen to advance step by step, and to lift up and undermine the protecting layer of tin. This metal protects only the part it covers, but it renders the iron more liable to rapid rust than if it were not there.

ALUM BATTERY.

Alum or potassio-aluminic sulphate ($KAl_2\ 4SO_4$) is employed in several branches of industry.

A German physicist arranged a battery whose electrodes were ordinary zinc and carbon, and whose liquid was a solution of alum; this battery undergoes polarization, of course, but it is said to depolarize after the circuit has been open but a very short time.

The chemical action must be very complicated in this battery. When the already complex nature of the alum is taken into consideration, the addition of the zinc cannot but lead one to believe that the composition thus produced is extremely complicated.

At Mulhouse they use for electric-clock purposes a battery whose liquid is a mixture of sea-salt (500 grammes) and pulverized alum (200 grammes) dissolved in water. This application deserves notice, as constant batteries are generally thought to be necessary for the service of electric clocks.

The cells of the above battery are very large: the hollow carbon cylinder has an exterior diameter of $4\frac{3}{4}$ inches and an interior diameter of $3\frac{1}{2}$ inches; the plate of zinc is $2\frac{2}{3}$ inches wide, and these two electrodes are immersed about 10 inches in the liquid. There are twenty clocks distributed in two distinct circuits; there are two closings of the circuit a minute, each lasting one second. There are sixteen cells, of which two are charged every week in order to always keep the battery the same; each cell works, therefore, four months without any attention whatever.

Another step in advance has been made in the arrangement of a battery which is only renewed once in every two years; it is, however, only used in the fire-telegraph service, which demands but little work.

These practical examples show once more that if the use of single-liquid batteries is well understood, they can be employed in many instances.

REMARKS UPON SINGLE-LIQUID BATTERIES.

It might be generally said that by taking any two pieces of different metals, or a piece of metal and a piece of carbon, and immersing them in some liquid conductor of electricity, a battery could be made.

The more lively the action of the liquid upon the positive electrode (negative pole), the greater intensity the battery will possess; there will be no action upon the other electrodes, at least not during the passage of the current; that is, not while the exterior circuit is closed. The choice of this second electrode is, however, far from being a thing of indifference; the less the electrode is capable of being attacked by the liquid, the greater will be the intensity of the battery. That is the reason why platinum and carbon should be preferred, at least with sulphuric acid, nitric acid, and the other liquids of which we have spoken.

The electric action is the result, in reality, of the difference of two chemical actions, one of which takes place while the other is prevented; one of the electrodes is attacked, the other is preserved from the attack of the liquid (at least while the circuit is closed).

The more energetic the action upon the positive electrode and the less this action upon the negative electrode, the more developed will be the electric phenomenon. For instance, a zinc-acidulated water-iron battery has but little power; the action of the liquid upon both electrodes is very lively as long as the circuit is open; as soon as it is closed the action upon the iron is stopped; but it reacts upon the attack of the zinc and diminishes it. If platinum be substituted for iron, the zinc alone is acted upon and the platinum remains unattacked before the circuit is closed; when the circuit is closed and the current flows, the action upon the zinc is hardly diminished.

Many persons have proposed compound liquids: mixtures of sulphuric acid and sea-salt, mixtures of salted water and flower of sulphur, etc. Satisfactory results may possibly be obtained in this manner; but this study has lost a great deal of interest on account of the invention of constant batteries, of which we will now speak.

PART II.

TWO-LIQUID BATTERIES.

INTRODUCTION.

We have already said that in order to successfully oppose polarization of electrodes, chemical substances capable of absorbing the hydrogen as it is given off upon the negative electrode must be employed. A second liquid is most frequently used for this purpose; nitric acid is eminently suitable for this office.

Experiment.—Let us take a zinc-sulphuric-acid-platinum battery; cause the current it furnishes to pass in a galvanometer. The deflection of the galvanometric needle is seen to decrease, and thus mark the progress of polarization. Let us now throw a few drops of nitric acid around the platinum, and the intensity of the current will be seen to increase immediately, thus making manifest a decrease in the polarization. It is easily understood that the nitric acid is decomposed by its contact with the hydrogen; water and bioxide of nitrogen are formed, the freeing of which produces, at the contact with the air, nitric-tetroxide vapors, very sensible to the smell.

In order to use nitric acid most advantageously, it should not be spread throughout the whole mass of the liquid, but should be concentrated around the electrode to be polarized. To fulfil this condition porous jars have been adopted to separate the two liquids, one of which is designed to dissolve the zinc and the other to dissolve the hydrogen given off, or on the point of being given off, upon the negative electrode.

The denomination "two-liquid battery" is badly chosen, because in many instances, which we will cite, solids are used as depolarizing agents instead of liquids; it would have been more exact to say *two-electrolyte battery*, but as this appellation is not in use, we do not think best to adopt it.

In most cases chemical depolarization is obtained by means of substances capable of furnishing oxygen, which, combining with the oxygen, prevent the latter from freeing itself and polarizing the negative electrode.

In the experiment that we have described above, the nitric acid, being decomposed, produces oxygen, and hence the result. Nitric acid NO_5, being very rich in oxygen and easily decomposed, was naturally fixed upon; it is one of the best batteries known.

Other acids besides nitric acid could be used: chloric acid ClO_5, chromic acid CrO_3, permanganate acid Mn_2O_7 are indicated; but they are not generally used in this way.

Salts (chlorate of potash, bichromate of potash, etc.) are generally substituted, which give off oxygen under the influence of the sulphuric acid.

It may be said, in a general way, that all means of producing oxygen can be satisfactorily employed for the depolarization of the negative electrode of a cell.

Oxides from which oxygen is readily freed might be employed instead of acids; oxygenized water would be excellent if the difficulty in preparing and preserving it did not render it practically impossible; but the bioxide of manganese and the bioxide of lead may be used, as will be seen farther on.

We have said how that acids rich in oxygen were not used alone; they are generally in the shape of some salt from which they are freed by sulphuric acid, or, if need be, by some other. In the same manner, the combination of peroxides and sulphuric acid may be employed to give off oxygen.

All the processes which we have just described consist in the use of oxidizing bodies or oxidizing mixtures; there is one more kind to be indicated; namely, the use of chlorine, which is also an oxidant in the presence of water, because it tends to combine with the hydrogen and to free the oxygen.

All the means for producing chlorine may be used for depolarization.

We have yet to speak of a chemical means of depolarization very different from any that we have hitherto mentioned, and which according to many is the best. It consists in the use of salts, such as sulphate of copper, which decompose under the influence of the current, depositing their metal and checking the freeing of hydrogen. The result is that a metal is deposited upon the negative electrode instead of gaseous hydrogen; if, at the start, the electrode was of copper, its surface would remain unchanged, and consequently there is no polarization.

We will examine in detail all these means of depolarization and indicate the most important applications that have been made, always following the same method that we employed in the study of single-liquid batteries.

The first idea upon the processes which we have just mentioned dates from the year 1829, and is found in a memoir of Becquerel.

"It should be observed," says Becquerel, "that the battery carries in itself the cause of the continual diminutions in the intensity of the electric current; for as soon as it begins to work, there take place decompositions and transportations which polarize the plates in such a manner as to produce currents contrary to the first one. The art consists, therefore, in dissolving these deposits, as they form, by means of properly placed liquids. . . . This is attained by means of the process that I have described. . . . By thus diminishing the intensity of the secondary current, sensibly constant effects may be obtained."

It is this view, so plainly expressed fifty years ago, that has suggested the present work and which justifies the classification that we have adopted.

The first indication of a battery depolarized by means of a salt of the metal which constitutes the conducting electrode is found in the following words taken from Becquerel's memoir:

"Let us continue to use saturated solutions of metallic salts, which cause no decomposition of the immersed metal. Let us then put with the copper a saturated solution of nitrate of copper, and with the zinc a saturated solution of sulphate of zinc. The deflection of the galvanometric needle will reach 88° and then undergo but a slow diminution. An addition of nitric acid to the solution of nitrate does not modify the intensity of the current. The result is the same when sulphuric acid is added in the solution of sulphate, the zinc having been well cleaned. Here is then a maximum effect."

Finally, Becquerel speaks of depolarization obtained by means of nitric acid; he experiments upon a cell containing zinc, copper, and a porous partition, the common liquid being a saturated solution of sulphate of zinc. He says:

"According to the general rule the zinc ought to be more attacked than the copper, and such is the result; the deflection is then 62°, and if a few drops of nitric acid be added in the compartment containing the copper plate, where the chemical action is most feeble, the galvanometric needle will mark 86° and will remain stationary for some time. . . . The same quantity of acid put with the zinc will sensibly diminish the intensity of the current." A little farther on he says: "I once succeeded in obtaining a compensation, so that the needle's deflections remained constant during one hour, an advantage never found in ordinary batteries."

The only thing that escaped Becquerel's notice is the part that the hydrogen plays in polarization; he did not observe the nature of the chemical reactions which take place in those batteries now termed Daniell's battery and Grove's battery. He studied batteries from a purely physical stand-point and neglected the chemical problem.

CHAPTER I.

DANIELL'S BATTERY.

IN a first series we will put all the batteries in which depolarization is effected by the use of salts. In order to facilitate our exposition, we ought to divide this series into several categories, that of sulphates, that of chlorides, etc. In fact, the two that we have just mentioned are the only ones which possess any importance up to the present date.

DESCRIPTION.

As we are not paying any attention to the chronological order of discoveries, we will not describe the battery

FIG. 21

of Daniell under the form given to it by its inventor in 1836; the form which we give is that in present use.

Fig. 21 represents three cells joined in intensity. In

DANIELL'S BATTERY.

each are seen the outside glass jar, a thin hollow cylinder of zinc, z, a porcelain porous jar, and a strip of copper, c.

The two liquids—a saturated solution of sulphate of copper in the porous jar and dilute sulphuric acid in the outside jar—are separated, but communicate with each other through the pores of the porcelain jar. The copper electrode is immersed in the sulphate, and the zinc in the acidulated water.

The only difference between this battery and that of Volta is the addition of sulphate of copper around the copper electrode. The zinc dissolves, oxidizing and forming sulphate of zinc. The hydrogen produced by this reaction, instead of being given off upon the negative electrode, takes the place in the sulphate of copper of an equivalent quantity of copper, which is deposited upon the electrode. This deposit not changing chemically the surface of the electrode, it is plain that there is nothing like polarization produced. In other words, the addition of sulphate of copper is sufficient to completely depolarize the negative or conducting electrode.

Such is the simple combination due to Daniell, the most perfect as yet invented.

If the action of the battery continues for some time, all the sulphuric acid in the outside jar will be converted into sulphate of zinc; the action, however, is not in the least checked by this, and the electro-motive force remains almost the same. The chemical action that then takes place consists in the substitution of zinc for the copper in the sulphate of copper, and the progressive transformation of sulphate of copper into sulphate of zinc.

In general practice the battery is not charged with dilute sulphuric acid; neither is any sulphate of zinc put in the outside jar, but simply pure water. At first the action is more feeble, and the internal resistance of the cell much greater; but the sulphate of copper, which traverses the porous partition, is transformed into sulphate of zinc by the action of the zinc, and the pure water is soon found to contain a certain proportion of salt, by which its conductivity is increased. This first period lasts a longer or shorter time, according to the circumstances; but a very effectual way of shortening it consists in closing the circuit with a very short conductor, which has no resistance; the chemical action is thus made more lively, and at the end of an hour or two the battery may be said to have reached its normal state of work.

Electro-motive Force.—As this battery is not polarized, its electro-motive force ought to be invariable; in fact, the two expressions are synonymous.

The experiment shows that the electro-motive force of Daniell's battery is indeed very constant. In the practice it may be taken as a unit, and others can be compared with it.

The British Association has adopted a unit differing very little from this one, and has given to it the name of Volt. The cell, in which the electro-motive force is exactly equal to the volt, differs but slightly from that of Daniell. It is a cell in which the copper is immersed in a solution of nitrate of copper, and the zinc amalgamated in sulphuric acid diluted with twelve times its weight of water.

The electro-motive force of Daniell's cell is, we said, very constant, and it varies but slightly with the temperature.

It has been found that if it is 1000 at 18° centigrade, it will only reach 1015 at 100° centigrade. It changes

DANIELL'S BATTERY.

very little with the acidity of the liquid. If it is 1079 with acid diluted with four times its weight of water, it only diminishes to 0.978 with acid diluted with twelve times its weight of water. The richness of a solution of sulphate of copper has but little influence upon it.

M. Regnauld has given the following figures concerning a cell without sulphuric acid and charged with solutions of sulphate of zinc and sulphate of copper:

Solution saturated with sulphate of copper...................... 175
The same, diluted with twice its bulk of water................ 175
The same, diluted with ten times its bulk of water... 174
The same, diluted with fifty times its bulk of water........... 172

These numbers are not expressed in volts, but a unit was taken which is equal to the electro-motive force of a bismuth-copper thermo-electric cell whose solderings are 0° and 100°.

The electro-motive force of a Daniell cell with sulphuric acid and a saturated solution of sulphate of copper is equal to 179 of the above units. It is therefore seen that the substitution of sulphate of zinc for dilute sulphuric acid does not notably change the electro-motive force.

We have said that it did not vary with the richness of the solution of sulphate of zinc; it has been shown that there is no perceptible change when the solution is at first concentrated and then diluted with one hundred times its bulk of water. The thickness or nature of the porous jar has no influence; porous partitions of many kinds have been tried, such as gold-beaters' skin, pipe-clay, under-baked porcelain, tubes of rosewood and of the pear-tree, of ebony and of boxwood.

Resistance.—It must, however, be acknowledged that

the condition and force of a battery are always variable; for if at the start the solution is too weak, it becomes more and more concentrated, and its conductivity is constantly changing. The following figures show that it reaches a maximum and then decreases:

Solution concentrated with sulphate of zinc (specific gravity 1.441) 5.77
The same, diluted with its bulk of water.................... 7.13
The same, diluted with three times its bulk of water......... 5.43*

The conductivity of liquids varies also with the temperature; that of a solution of sulphate of zinc reaches its maximum at 14° cent.

If the nature of the liquid in the outside jar changes, that in the porous jar will also change. In fact, the solution of sulphate of copper gradually weakens, and its conductivity changes with its state of concentration and the temperature.

A few crystals of sulphate of copper may be added in the porous jar to keep up the solution. The more concentrated the solution becomes the heavier it gets, and in order to keep it in a state of saturation up to the top of the jar, they used to suspend crystals to the upper part by means of a small diaphragm of copper soldered to the strip of copper; but this precaution has been generally abandoned, and the crystals are now simply thrown in the bottom of the jar. This suppression causes no inconveniences, because the battery loses none of its qualities when the liquid ceases to be saturated; it is indeed an advantage, for the consumption of sulphate is less active.

A saturated solution of sulphate of copper has, how-

---

* See table 3, end of volume.

ever, a greater conductivity than when it is diluted with water, as the following table shows:

Sulphate of copper. Saturated solution (specific gravity 1.171) 5.42
The same, diluted to half.................................. 3.47
The same, diluted to quarter................................ 2.08

It should therefore be admitted that the use of a diluted solution increases the resistance of the battery; but it is probable that the sulphate of zinc in mixing with the sulphate of copper increases the conductivity of the liquid.

Endosmose causes the liquid in the porous jar to rise slightly above that in the outside jar; finally, evaporation causes the two liquids to gradually descend and increases the richness of the solution.

All these reasons, and others which we will soon give, go to prove that the resistance of Daniell's battery is constantly changing; every time it is measured it is found to be different.

The intensity of the current of Daniell's battery, as has been shown, varies considerably from one day or from one hour to another, and as the electro-motive force does not change, it is clear that it must be the resistance which varies.

Figures relating to the resistance of battery cells would possess but little interest, as they can only be approximate, and applicable only to jars and electrodes of determined dimensions, and to certain heights of the liquids in the jars.

SELECTION OF CELLS.

In general, larger cells are used for local circuits, and smaller ones for telegraph lines, because the former are less resistant than the latter; this difference results from the fact that the immersed surface of the electrodes is greater in the large ones, whereas their distance is almost the same.

If things were looked at from a physical point of view only, the use of large cells would always be advantageous; but in the practice there are other things to be taken into consideration. The convenience, and above all the economy in buying and keeping them in order, must be thought of. Large cells are more cumbersome and more exposed to accidents; besides, they are especially dear, and the necessary supply of sulphate of copper and zinc plates tends to increase the expense.

Consequently the use of cells as small as possible is recommended. It is necessary to understand perfectly how in long circuits resisting batteries present but few inconveniences, and how the larger cells are preferable in short circuits.

If in a very long circuit, of 3000 units of resistance, for instance, with a receiving instrument of 1000 units, a battery of 25 cells of 10 units each be employed, the total resistance of the battery will be 250 units, and that of the whole circuit 4250. The resistance of the battery is thus but a small fraction of that of the whole circuit; consequently the substitution of smaller cells, even if their resistance were twice as great as that of the first ones, would not greatly increase the resistance of the circuit, nor, consequently, the intensity of the current. But if, on the other hand, we take a short circuit of 2 units with a receiving instrument of 3 units, and use a battery of 5 cells, each having 10 units of resistance, there will be in the battery 50 units and in the circuit 55 units of resistance.

DANIELL'S BATTERY.

It is seen that the resistance of a battery is the most important element, and by diminishing it one half by the substitution of larger cells the resistance of the circuit will be reduced to 30 units, and consequently the intensity of the current will almost be doubled. Let us follow up this idea. The intensity of the current being much greater with the 5 new cells of 5 units of resistance each, their number might be reduced to 3, for instance; the resistance of the battery would then be 15, that of the circuit 20, and finally the intensity of the current ($\frac{3}{20}$ or $\frac{1}{8}$) would be still greater than that we had at the beginning. Thus for a short circuit larger cells should be used, only less in number, by which a saving is obtained.

POROUS JARS.—The porous partitions which separate the liquids in constant batteries are generally made of porous porcelain; but one can use wood, carbon, bladders, canvas, pipe-clay, paper pulp, and in general all substances not chemically acted upon by the liquids.

In his first experiments Daniell used a piece of bladder inside of and having the shape of a copper tube pierced with holes, which served as the negative electrode of his cell; this membrane partition presented great advantages, its cylindrical form was excellent, and it was very generally adopted; it was very porous and electrically but little resistant. It was abandoned because it was too porous and too fragile.

Finally porous jars, properly so called, were brought into use; they are made of porous porcelain. We will here only speak of this latter; those made of parchment paper, canvas, etc., will be treated of in the description of those batteries in which they are used.

We have explained how, in the regular and theoretical action of Daniell's cell, copper is deposited upon the copper electrode. In the practice it is found that copper is also deposited upon the inside surface of the porous jar, and even in the pores, which are finally stopped up; after a certain time the jar ceases to be porous enough and should be replaced. That is one of the inconveniences of this form of Daniell's cell, which, however, should not be exaggerated, as the porous jars are very cheap and may be replaced at very little expense.

In the telegraph service, a porous jar may sometimes serve six months or a year. Purchasers of old metals buy these porous jars, often very heavily charged with copper, and know how to use them to the greatest advantage.

These deposits of copper in the interior of the porous jar frequently present a very interesting peculiarity: they show themselves upon the surfaces with a tree-like aspect; it is a slow crystallization, which reminds one of the unfolding of a fern.

It frequently happens that the deposit of copper accumulates upon certain points of the inside surface, and presents a crystalline structure, properly so called. Crystals are sometimes seen to attain the dimensions of $\frac{2}{25}$ or $\frac{3}{25}$ of an inch.

These deposits in the pores and upon the surface of the porous jar may easily be accounted for. They are the result of an electro-chemical action analogous to that shown in a previous experiment by De La Rive upon ordinary commercial zinc compared with chemically pure zinc. The porcelain contains various heterogeneous particles, which produce local actions and also the decomposition of the sulphate of copper. It is probable that this deposit takes place very slowly at first, but as the deposit accumulates the process becomes more and more rapid.

These deposits of copper in and upon the walls of the porous partition possess another quality which we should not forget; namely, that of diminishing the internal resistance of the cell. This is a fact that oft-repeated experiments have proved. The explanation seems to us very clear. The porous partition presents a very great electric resistance, because the total section of the canals filled with liquid is extremely small. When a portion of this section is replaced by copper, a metal possessing great conductivity, the resistance should be reduced in proportion.

It should be noted that, under these circumstances, the copper causes no polarization of the electrodes, otherwise the results would be entirely different.

All this would lead to the belief that the choice of porous jars had great influence upon the intensity of the current, by changing the internal resistance of the cell. The results of several experiments show, however, that jars of different states of porosity can change but slightly the resistance of cells.

The conclusion points to the preference of jars of little porosity in all batteries destined for continual and prolonged work. They should be immersed in water before charging the battery; or better, charge the battery several hours before using it, in order to allow the liquids to penetrate the pores and come into contact with each other.

Very porous jars have the disadvantage of allowing the liquids to mix too easily; the result is a direct action of the zinc upon the sulphate of copper, and a deposit of the copper upon the zinc, which is a grave fault inherent in Daniell's battery, and upon which we will now enlarge.

LOCAL ACTIONS; WASTE.—The sulphate of copper,

which penetrates the porous partition, comes into contact with the zinc, and is decomposed by an electro-chemical local action. Copper is deposited upon the zinc in the shape of black mud, which cannot be put to any use. (This black powder is said by some to be oxide of copper, but this is an error, as can be easily shown by means of some feeble acid, as acetic acid.)

This action does not co-operate in the production of the current; it takes place especially, as can be easily understood, with very porous jars; but it exists in every arrangement of Daniell's cell, as will be seen as we advance.

This inconvenience would be less if the current circulated continually; but in almost all applications, batteries, and especially those of Daniell, are only used intermittently and at long intervals. Many of the telegraph batteries and those used for electric bells work but a few minutes daily, thus being nearly always in an open circuit. Under these circumstances, the fault of which we speak is at its maximum.

It is understood that if, on the contrary, the cells are very actively employed, as in an important telegraph office, this same fault is less grave; the sulphate of copper and the zinc are consumed much more usefully. But even in that exceptional case, where a Daniell cell furnishes a constant current, there is a certain quantity of sulphate of copper which, penetrating the porous jar, is decomposed by its contact with the zinc, and deposits upon the latter, copper in the shape of black mud.

In short, we can say that an ordinary Daniell cell consumes almost as much zinc and sulphate of copper when its circuit is open, and when it does no useful work, as when its circuit is closed, and when it really acts as a producer of electricity.

This is about the only fault that can be found in the Daniell battery. Its gravity, however, has often led to the preference, in many instances, of other batteries, such as that of Leclanché, which, however, do not fully fill the conditions of a constant battery.

This fault is present in the Daniell battery to a greater or less degree, according to the porosity of the jars. Among the improvements upon Daniell's battery, which we will soon describe, there are some which possess the advantage of greatly reducing the detrimental effect in question.

The copper, by entering the pores, diminishes the porosity, thereby improving the battery, for a certain length of time at least. But it is plain that, if the porosity be too much reduced, the battery will no longer work, for if the pores of the partition are completely stopped up, it would no longer furnish any current.

Finally, the pure loss in the consumption of sulphate of copper is less when the solution of sulphate is less concentrated; and, as the reduction of concentration does not diminish the electro-motive force, and does not notably increase the resistance, there is a great advantage, as can be seen, in not using a saturated solution of sulphate of copper.

It should be noticed that the deposit of copper upon the zinc does not change the electro-motive force of the cell, which confirms what we said of the constancy of this force in Daniell's cell. In some models, the zinc is suspended, so that it may not rest on the bottom of the jar and come in contact with any rubbish that might happen to accumulate there, which would result in local actions. It is an established fact that, in ordinary batteries, the lower part of the zinc is more consumed than that near the top. The cause of this is due partly to the contact of this rubbish with the zinc, producing local actions.

CLIMBING SALTS.—The Daniell battery possesses one more little fault that should not be left unnoticed. When the liquid in the outside jar begins to become saturated with sulphate of zinc, climbing salts are formed, which ascend the walls of the outside jar, and even of the porous jar, running over the rim of the former and descending on the outside. All this can, however, be avoided by proper care and attention. These salts establish permanent communication between the different elements of the cell, which communication contributes to the waste. The manner in which these salts are formed is very easy to understand.

In the beginning, either by a slight movement or otherwise, some of the liquid is thrown upon the sides of the jar. Evaporation causes the disappearance of the water and leaves crystals. Immediately, capillary action sets in, either between the crystals and the side of the jar, or along the side of the crystals, and a small quantity of liquid thus rises above the general level.

Evaporation again causes the formation of new crystals, which is facilitated by the thinness of the layer. This action continues step by step horizontally and vertically.

As long as the formation of climbing salts does not run over the top of the glass jar, or establish any permanent communication between the elements, it is not unfavorable. It might indeed be considered as an advantage, as it reduces the concentration of the sulphate of zinc solution. We have previously shown by figures that a saturated solution has less conductivity than when diluted to half. When the solution is saturated, it is incapable of dissolving the salt formed by the action of the battery, and the most unfavorable thing that can then happen is a deposit of crystallized sulphate of zinc upon the zinc or upon the immersed part of the porous jar, for it stops the chemical action or the communication of the liquids. It is therefore advisable to take the climbing salts completely away and not to put them back in the glass jar.

To suppress the production of climbing salts, it has been proposed to spread a thin layer of oil upon the surface of the liquids, which would check all evaporation; this has, however, received no application, on account of the uncleanliness it would inevitably cause. This layer of oil would also occasion rapid concentration, which, as we have stated, diminishes the conductivity. The climbing salts may be kept from running over the top of the glass jar by smearing the top with some greasy substance (paraffin, for instance).

Climbing salts of sulphate of zinc are very easily detached from the glass, either with the fingers or with a rag; but they hold much more firmly to the porous jar, and it is with difficulty that they may be detached from it by rubbing. Therefore, the top part of the porous jar which stands out of the liquid should be carefully glazed; the salts may then be taken off as easily as from the glass.

IMPROVED DANIELL CELL.

We have sought to construct as satisfactory a Daniell element as possible. The following is the disposition that we have decided upon:

The positive electrode is formed of a cylinder of zinc surrounding the porous jar, and is held at a small distance from the latter by means of small sticks of wood placed vertically between the two; the zinc and the small sticks are held in their place by being tightly bound together at the top and bottom with pieces of string.

The connecting strap of the zinc is cut out of the same sheet of metal as the cylinder itself, which dispenses with any loss in the cutting, provided two cylinders with their connecting straps be cut out of the same piece, one in the reverse manner from the other.

The copper electrode is cut in the same manner out of a sheet in the form of a cylinder, which is placed inside the porous jar; small sticks of wood placed around the copper and bound to it with string keep it from coming into contact with the interior surface of the porous jar.

At two thirds of the height of the copper is fixed, without solderings and simply fastened in the copper, a circular piece of copper pierced with holes, thus forming a partition without impeding the movement of the liquid. Upon this partition are placed sulphate of copper crystals, which hold the solution at saturation.

In the outer jar is put a solution of sulphate of zinc possessing its maximum conductibility; that is, a solution at first saturated and then diluted with its bulk of water (specific weight about 1.10). This disposition is intended to reduce to a minimum the internal resistance of the element, and to render it as compact and solid as possible. For experiments of precision it is best to ascertain, at first starting, the density of the sulphate of zinc by means of an hydrometer, and to always keep it the same by adding, from time to time, some pure water, as the solution becomes concentrated either by evaporation or by the formation of sulphate of zinc in the battery.

Undoubtedly the use of a saturated solution of sulphate of copper greatly increases the expense of the battery, but we are now supposed to be talking of an apparatus designed for experiments of precision and in which the question of expense is secondary.

From this point of view the precautions that we have suggested appear to us to be indispensable. How indeed could the internal resistance of a cell be advantageously measured unless the respective distances between the electrodes were invariable and the composition of the liquids known and determined?

It is only with elements arranged in the above manner that it may be hoped to obtain concordant or even comparable results. In truth, it is very probable that these precautions will not be sufficient, but they are necessary.

BALLOON BATTERY.

Another arrangement of Daniell's battery, represented by Fig. 22, has been and is still used in some countries. It needs no attention for six months, or more, at a time. The flask, which surmounts the cell, contains two pounds of sulphate of copper crystals, and is filled with water. The flask is closed with a perforated cork fitted with a glass tube; this tube descends as far as the liquid in the porous jar. The solution of sulphate of copper being more dense in proportion as it is more concentrated, it can be seen that the part of this solution which is in the porous jar is constantly held at saturation, for as it weakens it is supplied by the saturated solution which descends from the flask.

The glass jar of the cell is closed with a wooden lid, which supports the flask. The result is that evaporation is reduced to almost nothing, and, consequently, there is a very slight or no formation at all of climbing salts, and the liquid is preserved.

TWO-LIQUID BATTERIES.

This battery has been known to work for more than a year without requiring any attention, during which time it satisfactorily met all practical requirements.

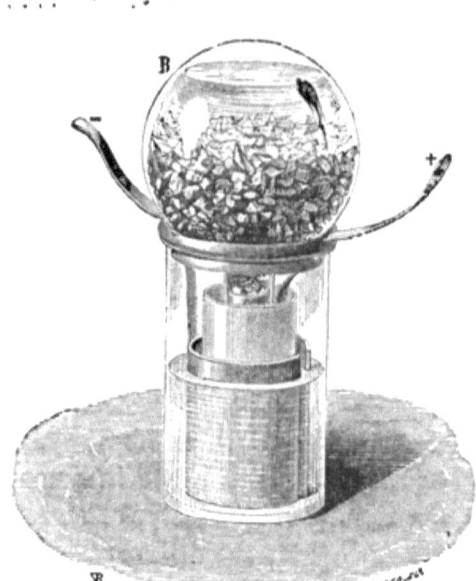

Fig. 22.

It has, however, been replaced by more economical batteries, such as Callaud's or Leclanché's.

A REVERSED FORM OF DANIELL'S BATTERY.

For a long time Daniell's battery was arranged in a manner the reverse of that which we have already described. The zinc, in the shape of a solid cylinder, which served as soluble electrode, was placed inside of the porous jar, instead of outside. The copper was placed around the outside of the porous jar, serving at the same time as conducting electrode and as the jar containing the liquid. In addition to the exterior hollow copper cylinder, another was placed nearer the porous jar, in order to diminish the resistance of the cell. This interior cylinder of copper was pierced with holes, to facilitate the circulation of the liquid (solution of copper sulphate). Finally, crystals of copper sulphate were put between the two copper cylinders, in order to keep the solution in a state of saturation, in spite of the consumption of the battery.

At first sight, this arrangement would appear to be much superior to that previously described. In this one the glass jar is suppressed, and consequently the many accidents resulting from its brittleness are avoided. It has, however, been abandoned for the first arrangement on account of the following reasons:

1st. The battery with the copper jar costs a great deal more, because of the large quantity of copper required, and also on account of the quantity of copper sulphate with which the battery is charged at the beginning.

2d. The copper jar might become pierced, and the liquid would then leak out; a few impurities in the metal would suffice to set up local electro-chemical actions, and thus bring about perforation.

3d. Cast zinc is used, which presents another disadvantage. For, during the process of casting, very often little cavities are formed in the zinc, into which the liquid penetrates, thereby producing local actions and uselessly consuming the zinc.

4th. The zinc nearly fills the porous jar, thus leaving but little room for the liquid, which is soon saturated with the sulphate of zinc and becomes incapable of dissolving any more. This is a grave fault in the working of the battery.

We must do Daniell the justice to say that he had foreseen this difficulty, and had proposed an accessory arrangement for the renewal of the water destined to dissolve the sulphate of zinc; but this addition complicated the apparatus and increased the cost.

The cell might be sensibly improved by the substitution of a thin, hollow cylinder of zinc for the massive zinc used above; the quantity of water in which the zinc is immersed would thus be greatly increased.

It is well to note, before leaving this subject, that the surface of the negative electrode is comparatively much larger than that of the soluble electrode. We said, in speaking of Wollaston's battery, that this condition was very favorable in single-liquid batteries, but it is not so in two-liquid batteries, or at least not in Daniell's battery. Since the depolarization of the negative electrode is complete, there is no advantage in increasing its surface. The considerations which prevail in the choice of electrodes have been clearly indicated in that which we have just said.

TROUGH BATTERY.

Still another arrangement of Daniell's battery is represented by Fig. 23, and the above name given to it.

A trough is made of teak and divided into ten cells by slate partitions; each cell is then subdivided by a porous partition of unglazed porcelain. A zinc plate is placed in one of these divisions, and a thin copper plate in the next one, and so on, until the ten cells are occupied. The copper plate of one cell is permanently connected with the zinc of the next cell by a copper strap cast into the zinc and riveted to the copper, which is easily bent over the slate partition.

DANIELL'S BATTERY.

The last copper and the last zinc plate are each connected to brass binding screws or terminals, which become respectively the positive and negative poles of the battery.

A solution of sulphate of copper and a few crystals are placed in the copper divisions; in the others, pure water or a very weak solution of sulphate of zinc.

This arrangement presents great advantages; it dispenses with glass jars, which sometimes break without any apparent cause. The trough is made water-tight by coating it internally with marine glue, and the liquids ought not to leak out. But it unfortunately happens, sometimes, that the marine glue chips off, and one cell

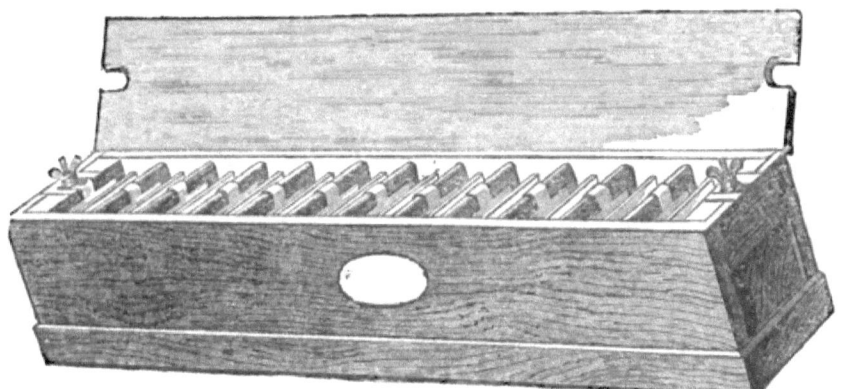

FIG. 23.

becomes leaky. When this occurs, the battery must be repaired.

The trough is very solid, and is easily transported when not charged with liquid. The zinc and copper electrodes of each cell are at a well-regulated distance from each other, and do not touch the porous partition if the battery is carefully charged.

These last conditions are fulfilled with difficulty where cylindrical elements are used. The trough having a wooden lid, there is very little evaporation. Of all known forms, this is the least cumbersome.

The dimensions of the electrodes are generally, for the zinc, $3\frac{1}{2}$ in. by 2 in., and for the copper, 3 in. square. The battery will work a month without the necessity of opening the trough. One of these batteries of ten cells costs $5.25, and the keeping it in order $2 per annum.

CONVENTIONAL FIGURE.

Batteries are generally represented by a conventional figure, which originally represented Daniell's trough bat-

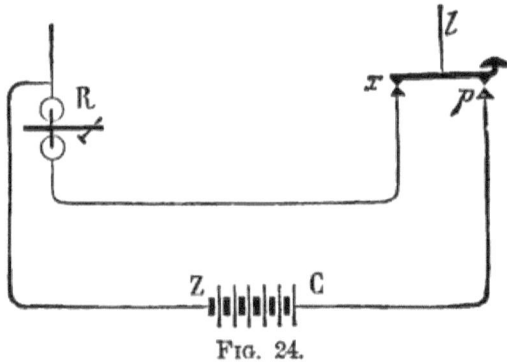

Fig. 24.

tery, or the sand battery, of which we spoke in Part I. Each cell is represented by two lines (Fig. 24), the short and thick one representing the zinc, and the long and narrow one the copper. Z and C mark respectively the negative and positive poles of the battery.

MUIRHEAD'S BATTERY.

There are a great many such in use in England
The outside jar is made of white porcelain and is

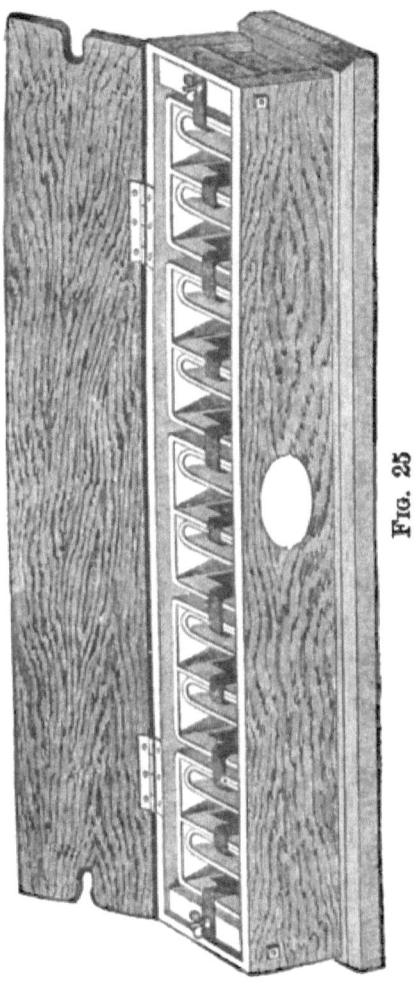

Fig. 25

square. The porous jar, made of red earthenware, contains the negative electrode and the sulphate of copper, and is placed in the square porcelain jar, which contains the zinc electrode and the sulphate of zinc. The electrodes are the same as in the preceding battery.

For economical reasons these cells are taken by twos; that is, each outside porcelain jar contains two compartments and two complete cells. This arrangement presents a very favorable condition, which is also met with in the cylindrical cells described at the beginning; viz., the compartment containing the sulphate of zinc is quite large. Fig. 25 represents several of these cells together.

CARRÉ'S BATTERY.

Carré's battery differs from the ordinary Daniell battery simply in the substitution of a vessel made of parchment paper for the ordinary porous jar. This porous partition offered very little resistance, which realized the object of its inventor. The whole battery indeed was arranged with a view to diminishing the resistance. The zinc cylinders were 22 in. high and $4\frac{1}{8}$ in. in diameter.

Sixty of these cells were used by M. Carré for electric-light purposes, which is, we think, worthy of mention, as it was the first time that Daniell's battery had been tried in that way.

In fact, M. Carré's arrangement was only fit for electric-light purposes; that is, to furnish a continuous current of great intensity for several hours. The frailty of the porous partition rendered the battery useless for any work of long duration. We believe, however, that this battery, in spite of the disadvantage we have pointed out, should again be taken up by persons interested in the electric light, who could give it a fixed place and could take care of it. In the use of this battery, the disagreeable acid vapors, which are dangerous to inhale, would be avoided, and the expense would be comparatively small.

This battery has been known to work 200 successive hours without any sensible weakening, by carefully replacing, every 24 hours, a part of the sulphate of zinc with pure water.

SIEMENS AND HALSKE'S BATTERY.

This battery is a Daniell battery with a porous jar, like those which precede, and is very extensively employed in Europe.

The copper, c, is at the bottom of the glass jar (Fig. 26), and the diaphragm or porous jar has the form of a bell. A thing to be noted is the central chimney, in which there is a glass tube through which a copper wire, attached to the negative electrode, passes, forming the connection of the positive pole. The porous jar sustains a mass of paper pulp, dampened with sulphuric acid, and then dried. The zinc, z, placed on top of the pulp, is a very thick cylinder, melted in a mould and carrying a vertical appendix, to which the positive connection of the adjoining cell is attached.

The arrangement of this battery has been changed several times. At first there was no porcelain porous jar, but simply the paper pulp. The general characteristics, however, have remained the same. The great thickness of the porous jar suppresses almost completely the diffusion of the sulphate of copper, and consequently the waste chemical action and the useless consumption of zinc and sulphate of copper are avoided. On the other

110 TWO-LIQUID BATTERIES.

hand, however, the internal resistance of the cells is considerable, so that they cannot be used as local batteries.

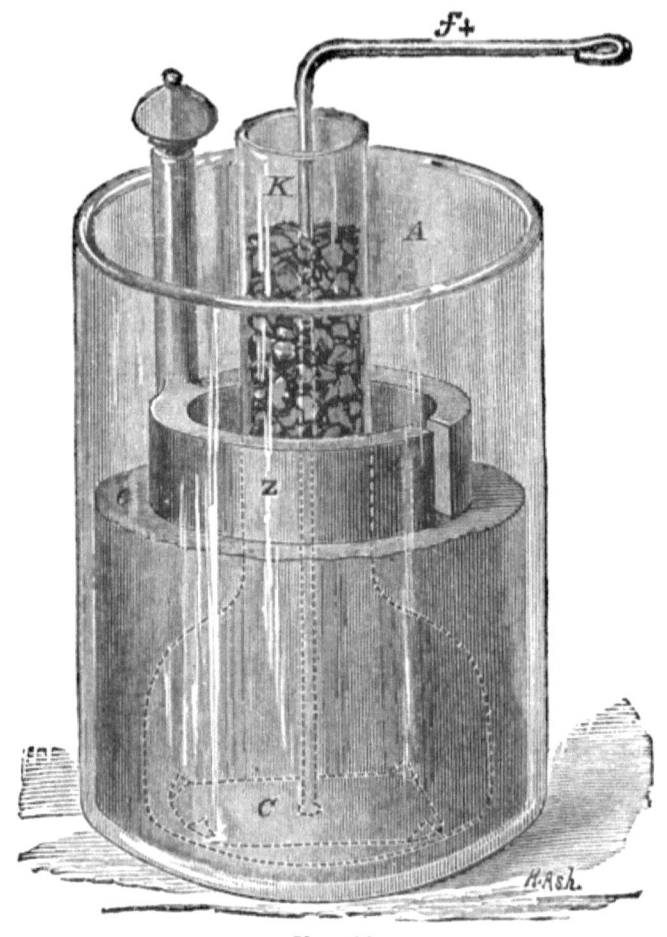

FIG. 26.

VARLEY'S BATTERY.

Cromwell Varley suggested a means of completely suppressing the passing of the sulphate of copper through the porous partition, a thing which Siemens and Halske only slightly diminished and slackened. This means consists in the substitution of oxide of zinc for the paper pulp in the preceding battery. It is easily understood that the sulphate of copper, which enters the mass of oxide of zinc, forms sulphate of zinc, and deposits oxide of copper in the shape of a black powder.

This original idea was perhaps never put into application outside of Varley's laboratory. But it deserves notice, for it shows the reader what an infinite variety of resources chemistry presents to those who know how to search for them.

It is clear that this porous partition becomes destroyed in time; it is also very resistant. These are all inconveniences in daily practice, but are unimportant in laboratory experiments, which require as perfect a battery as possible.

MINOTTO'S BATTERY.

This is in form a Daniell battery, extensively used in Italy and throughout British India. It consists of a jar, at the bottom of which is a copper plate fitted with a wire covered with gutta-percha, which ascends to the top of the jar and serves as a connection. This electrode is covered with an inch of crushed sulphate of copper, this again with a layer of river sand, and finally a plate of zinc of considerable thickness. The crushed sulphate of copper is separated from the sand by a piece of cloth or blotting-paper. Sir William Thomson recommends sawdust instead of sand. The zinc is of convex form, in order to permit the freeing of bubbles of hydrogen which sometimes form themselves in Daniell's battery.

These batteries last eighteen or twenty months upon important telegraph lines, and as long as thirty-two months upon less important lines.

TROUVÉ'S BLOTTING-PAPER BATTERY.

Trouvé's battery is one of the latest modifications of Daniell's battery. Fig. 27 represents a separate cell.

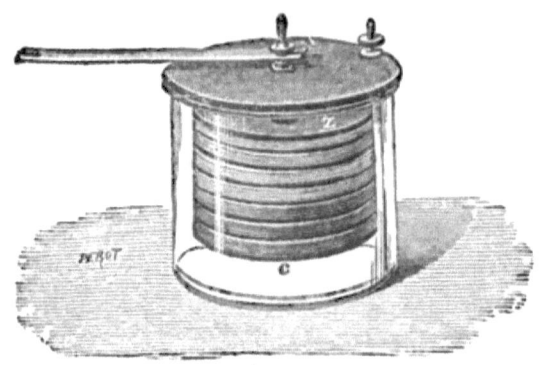

Fig. 27.

At the top there is a plate of zinc, at the bottom a plate of copper, and between the two a considerable thickness of blotting-paper. The upper half of the blotting-paper is wet down with a concentrated solution of sulphate of zinc, and the lower half with sulphate of copper. We have, therefore, all the elements of a Daniell battery.

It is of course plainly understood that when this battery is perfectly dry it is absolutely inert. To make a source of electricity, or, more simply, a voltaic cell, water must be added; only that quantity which the paper can absorb is necessary. Thus there is no free liquid, and Trouvé might have called his battery a *moist battery*, in order to distinguish it from liquid batteries properly so called.

Let us return to the description of the cell. The two electrodes and the blotting-paper are held by a central piece of copper, insulated by a tube of ebonite, which goes above the zinc and penetrates the cover of the glass jar, whose edges should be well ground in order to prevent all evaporation. This central piece of copper is finally furnished with a binding screw, to which the connection of the adjoining cell or of the circuit is fastened; the negative pole is a copper wire soldered to the upper part of the zinc electrode.

Several advantages of this battery are easily seen. When it is dry there is no waste whatever, although it be charged. To dry it, it is only necessary to expose it for a few hours to the sun or to a current of air. If it is to be used, the elements must only be moistened. If the cells are separated from each other like the one represented in the figure, they must be put under the faucet of a fountain which runs slowly. This operation should be repeated after a short interval, in order to allow the first water to saturate the paper to the centre. The cell is sufficiently moist when, by pressing the zinc and the copper between the thumb and index, drops of water are seen to ooze out upon the surface of the paper. When several cells are attached to the same cover, as in the military or medical batteries which we will describe farther on, they are immersed in special vessels and left there about a half-minute. Whatever the method be (and the choice is not of much importance) of moistening the elements, this done, they are placed in the glass jar or in the ebonite box, where they may remain several months, always ready for work, and indeed continually working and furnishing a remarkably regular current.

This regularity constitutes a new advantage in Trouvé's battery, upon which we will insist before passing to its applications.

The liquids of ordinary batteries have a movement but slightly impeded by the porous jar, which results in their mingling, as we have several times explained. The consequence is that local actions, which disturb the normal working of the battery, take place and occasion waste.

In Trouvé's arrangement the movement of the liquid is almost suppressed, consequently there is very little mixing of the sulphate of copper with the sulphate of zinc, and there are scarcely any local actions; that is, there is no reduction of the sulphate of copper upon the surface of the zinc. An evident economy of matter consumed by the battery, and other interesting facts, are the results. The liquids keep their respective positions almost without mixing, the resistance varies very slowly, and consequently the intensity of the current possesses great constancy.

We have often made the following experiment with a Daniell porous-jar battery. Having closed the circuit in the evening upon a galvanometer, it was found in the morning, twelve hours later, that the deflection was exactly the same. But if the circuit were opened only one minute and then again closed, there was observed a notable difference.

If the same experiment be made with Trouvé's battery, the intensity of the current is found to be the same, after the momentary rupture of the circuit, as it was before.

This difference in the results shows that the variation

of the resistance takes place suddenly and abruptly in the ordinary Daniell cell, and on the other hand very slowly in Trouvé's arrangement.

We have insisted upon the constancy of the electromotive force of Daniell's battery in general. Under the form given to it by Trouvé is added that very important quality which consists in the existence of a slightly variable resistance.

RECHARGE OF THE CELL.—When the cell has worked several months, more or less, the sulphate of copper is used up and converted into sulphate of zinc, and in order to give to the cell its first energy it must be recharged in the following manner:

The sulphate of zinc, which has taken the place of the sulphate of copper in the lower half of the paper, must be dissolved in pure water. For this purpose there are especial vessels having the desired level of the water marked upon their sides, for the sulphate of zinc which is in the upper half must not be dissolved. This same vessel must then be filled to the same level with a warm saturated solution of sulphate of copper, in which the lower half of the elements are immersed and left there three or four minutes. At the end of this time the sulphate will have penetrated from the circumference to the centre of the paper, and the cell is recharged. The cell should then be left to dry, and when it is to be used it should be dampened as we have said in the beginning.

MILITARY BATTERY.—Trouvé's battery has been applied in military telegraphing and with very good results. This battery consists of nine cells, distributed in three boxes, each containing three cells, as shown in Fig. 28. The boxes are made of ebonite, and have slate lids, to which the cells are attached as in the separate cell, Fig. 27. The three boxes are then placed one above another in the desired order, and enclosed in a large oaken box. This outside box is carried by means of a strap thrown

Fig. 28.

over the shoulder, or placed upon a wooden frame fastened on the back. The cells are $2\frac{3}{8}$ in. in diameter and $1\frac{1}{4}$ in. in thickness.

MEDICAL BATTERY.—For the medical application of a continuous current, Trouvé arranged a battery of three small cells, each being about an inch in diameter and $1\frac{1}{4}$ in. in thickness. The result shows a considerable resistance. These cells are taken by forties, sixties, or eighties, and attached to a single piece of slate, and placed in a water-tight wooden box, to the cover of which the connections are fastened, so that the whole or only part of the battery may be used at will.

These cells last a long time, because they contain a

comparatively great quantity of sulphate of copper, and because they work in circuits of very great resistance.

There is a very decided advantage in their great internal resistance for medical purposes; for the quantity of electricity which circulates is very small, and the electro-chemical decomposition at the contact of the conductors with the body of the patient is insensible.

RESISTANCE IN TROUVÉ'S BATTERY.—The largest model yet made is $3\frac{1}{2}$ in. in diameter and $2\frac{1}{8}$ in. in thickness. It is understood that, with this thickness, the resistance of the cell is inversely proportional to the section of the disc electrodes. The regular form of these cells is much more convenient than that of ordinary liquid batteries for exact calculations.

By modifying the diameter of the electrodes and the thickness of the blotting-paper between them, the resistance may be regulated at will.

CHAPTER II.

GRAVITY BATTERIES.

Several physicists became possessed, at the same time, of the idea of suppressing the porous jar, and of separating the liquids by their difference in density. Notable among these physicists were Callaud, Meidinger, and Varley. The last mentioned took out a patent before any of his competitors, in 1855.

Callaud's Battery.—This battery has been greatly improved since its first appearance. The following are the dimensions of the one most in use:

From the top of a glass jar 8 in. high is suspended a

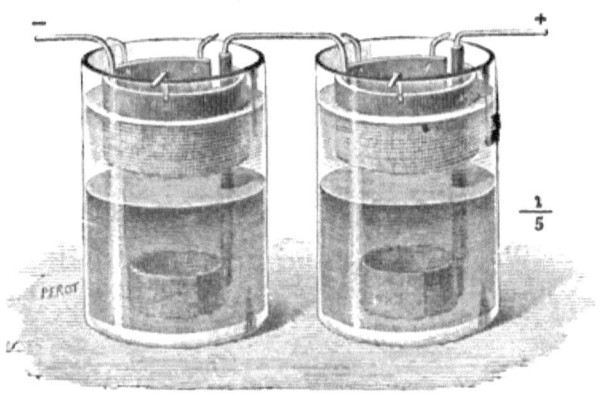

Fig. 29.

cylinder of zinc 2 in. high. It is held by three hooks riveted in the zinc and resting on the rim of the jar. The copper electrode is formed of a thin strip of copper

GRAVITY BATTERIES.

rolled in the shape of a cylinder 1¼ in. high and 1½ in. in diameter, and is placed at the bottom of the jar. A copper wire covered with gutta-percha and riveted to the copper cylinder passes through the liquid, and being twice bent, is soldered to the zinc of the adjoining cell. If the wire were not insulated by the gutta-percha, it would be liable to be cut at the line of separation of the two liquids.*

The solution of sulphate of copper is at the bottom of the jar, and remains there, because it is heavier than that of sulphate of zinc, which is placed above.

CHARGE AND MAINTENANCE.—The zinc being put in place, fill the jar to within half an inch of the top of the zinc with water containing one tenth of a saturated solution of sulphate of zinc. In general, when the solution of sulphate of zinc is poured in the water, the latter becomes slightly cloudy, which is the result of the presence of a small quantity of carbonate of lime in the water, especially if well-water; the addition of the sulphate of zinc brings about a reciprocal action; carbonate of zinc is formed, and sulphate of lime is precipitated. But this precipitate, being in a very small quantity, remains a long time in suspension, during which time the whole liquid assumes an opaline tint. At length the sulphate of lime falls to the bottom of the jar. It does not sensibly alter the action of the battery.

The sulphate of copper is added by means of a siphon. The solution, prepared beforehand, is conducted to the bottom of the jar and increased until the water is within

---

* This is explained by a local action, the formation of a cell between the two liquids and the copper wire. It will be seen farther on that a current can be produced under these circumstances, and the natural result is that the copper is dissolved in one of the liquids.

a quarter of an inch of the top of the zinc. The sulphate of copper does not remain separate at the bottom of the jar; it mingles with the lower part of the rest of the liquid, and consequently it is a diluted solution and not a saturated one, which surrounds the negative electrode. We have already said, in speaking of the porous-jar battery, that a saturated solution presents no advantage and only renders the local action more energetic.

The quantity of sulphate mentioned is sufficient for one month. A too great weakening of the solution may be recognized by the discoloration of the lower stratum of liquid.

When the solution of sulphate of copper is again added, it is better to take off about a quarter of an inch from the top of the water, unless evaporation has already lowered the water, in which case some pure water should be added, in order to reduce the concentration of the sulphate of zinc.

It is best to examine the battery every three months; but if it is properly taken care of from month to month, it need only be thoroughly cleaned once a year. When this is done, the deposits formed upon the surface of the zinc should be scratched off, the jars washed, and the liquids renewed as in the beginning.

We add that twenty-six batteries, eighteen cells in each, distributed in as many stations, caused an expense of fifteen cents yearly for each cell. This result, calculated upon three years' experience, has the value of practical information.

LOCAL ACTIONS AND LOST WORK.—Daniell's gravity battery is not free from that great fault that we have indicated, which consists in the local chemical actions which do not co-operate in the production of elec-

GRAVITY BATTERIES.

tricity, and which even take place when the circuit is open.

The two liquids, one above the other, would mingle extremely slowly if it were not for several causes, which we will examine. If a gravity cell be closely observed, the formation of gaseous bubbles at the contact of the zinc, and even upon the surface of the copper electrode, is seen, and these bubbles, freeing themselves from time to time, produce an agitation of the liquid. Indeed, two distinct currents in the liquid are seen, descending on one side and rising upon the other, which brings about a very slow mingling of the liquids. We will again have occasion to speak of these bubbles of gas and their effects; it suffices for the present to observe the movement they produce in the liquid. The result of this movement is that a small quantity of sulphate of copper ascends to the top. This salt decomposes at the surface of the zinc; copper is deposited, and an equivalent quantity of zinc is dissolved. The copper generally attaches itself to the lower part of the zinc, and frequently in the shape of pendants or stalactites, which hang below the zinc. The longer these pendants become the more rapidly they grow, and if they reach the level of the sulphate placed in the bottom of the jar, the battery becomes rapidly exhausted. It is, therefore, very advantageous to detach these stalactites, which can be done, when the battery is examined, by sharp taps on the zinc. The zinc should be in a perfectly horizontal position; for if one side were lower than the other, pendants would form themselves rapidly on the lower side.

RESISTANCE OF CALLAUD'S CELL.—There are two sizes of Callaud's cell in use, one of which is much larger than the other; but the separation of the electrodes balances the difference in the dimensions, and the two cells have sensibly the same resistance.

This resistance has been observed to vary from 37 units at the beginning, when the water contains very little sulphate of zinc, to $51\frac{1}{2}$ units after 23 days' work; that is, when a considerable quantity of sulphate of zinc has been formed. These figures confirm that which we have previously said of the Daniell gravity cell; no exact figures even for batteries of well-determined dimensions can be given; only vague limits between which the resistance varies can be fixed upon. At all events, Callaud's cell should be considered as possessing a very feeble resistance, and as well adapted for local as for long and resistant circuits.

APPLICATIONS OF CALLAUD'S BATTERY.

This battery was not very extensively used by the English, because they claimed that, to work well, it should not be moved at all. This, however, should not be exaggerated, as in France the battery is placed in a box upon rollers. The box remains under the table during work, and is only pulled out when the battery is to be examined or recharged. This movement has no sensible effect upon the working of the battery.

It has been adopted in Italy with a slight modification. The glass jar, instead of having the shape of the ordinary one, is compressed in the middle, thus dividing it into two compartments. The zinc rests upon the border of this partition. The diameter of the jar at the top and bottom is four inches, and that of the middle portion is only two inches. These dimensions would seem to increase the resistance, without presenting any counterbalancing advantages. The dimensions of the Italian battery are smaller than those of the French model; but such as it is, it has for several years given great satisfaction, as much on account of its constancy as on account of the economy and facility of keeping it in order.

This battery is very extensively employed in the United States. The monthly expense of keeping in order 600 Callaud cells, distributed in three batteries which supply ten circuits, is about $30.

There is another modification in use in the United States, to which Mr. Lockwood, its author, has given his name. The peculiarity of this form consists in the use of two flat helices as the copper electrode. One of these helices is placed at the bottom of the jar, and the other above the sulphate of copper. This disposition certainly renders the resistance of the battery less. Lockwood uses crystals of sulphate of copper, which, according to our view, uselessly increases the cost and the local actions.

TROUVÉ-CALLAUD BATTERY.

Trouvé arranged a Daniell gravity battery, which is extremely cheap. Trouvé had in view the application of his battery to medical purposes—it furnishes a continuous current—which is at present used in many hospitals and by many physicians. It could also be very well employed for other purposes. The glass jar is $4\frac{3}{4}$ in. high and $2\frac{3}{4}$ in. in diameter. The zinc is held upon the rim of the jar by being bent over in three places by means of pincers. The negative element consists of a copper wire in the shape of a flat helix, which rises in a glass tube to the top of the jar. The connection between the cells is made by means of a spiral spring at the end of the wire which is soldered to the zinc, and to which is fastened the wire forming the connection of the adjoining cell.

This battery, shown by Fig. 30, differs but slightly

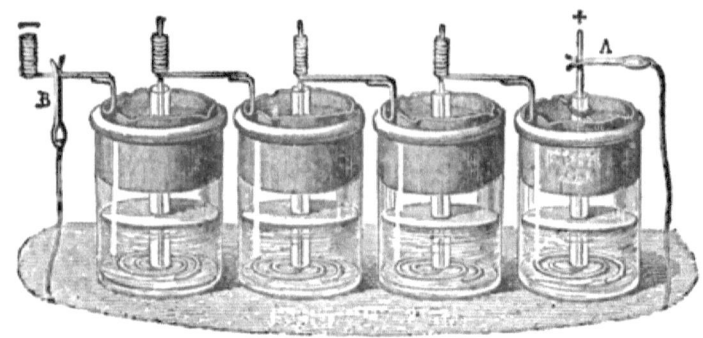

Fig. 30.

from the one we have described above; but it is very simple in arrangement, and on account of its small dimensions each cell costs only 12 cents.

MEIDINGER'S BATTERY.

This is one of the most extensively employed batteries in Germany. It has been tried in many other countries, but has been finally abandoned for the following reasons: Its cost is greater than ordinary Daniell or Callaud batteries; its internal resistance is also greater, and it occupies a considerable space.

The Meidinger battery (Fig. 31) consists of a large glass jar, A, at the bottom of which is placed a cup, d. In this latter is the negative element, formed of a thin leaf of copper, c, rolled in the shape of a cylinder, from which a copper wire rises in a gutta percha tube, g, to the top of the vessel, and forms the positive connection of the cell.

The cylinder of zinc, Z, is suspended above, as shown in

the cut, and descends a little below the top of the copper. The two jars are filled with liquid. The glass tube h, pierced with holes at the bottom, is placed in the centre of the cell, with the lower part in the cup; this central tube is filled with sulphate of copper crystals. The sulphate of zinc, being lighter, remains above the dissolved sulphate of copper. At the beginning, in order to increase the conductivity, magnesic sulphate is put in the

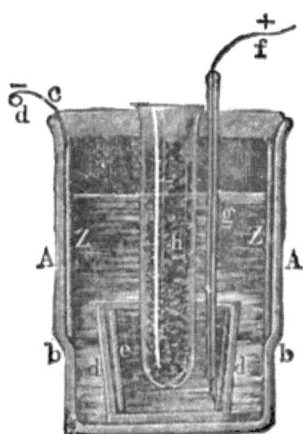

Fig. 31.

water, which solution has greater conductivity and is less dense than that of sulphate of copper.

The advantages of this battery are as follows:

1. It contains a large quantity of water which is required to dissolve the sulphate of zinc formed. In order that the battery may work a long time without any attention, this condition is very essential. We will see farther on that, when the solution of sulphate of zinc approaches saturation, it becomes more dense than that of copper, and that consequently the relative positions of the liquids are apt to change.

2. The deposits of copper which form upon the surface or hang in stalactites from the lower part of the zinc fall outside of the centre cup and thus do not touch the solution of sulphate of copper; this contributes greatly to the cleanliness and good condition of the battery. We have seen (in speaking of Callaud's battery) that, when these stalactites touch the solution of sulphate of copper, local action increases rapidly and assumes a very injurious intensity. This latter inconvenience is avoided in Meidinger's battery.

3. The cover which closes the cell and which is necessary to support the glass tube (sulphate of copper reservoir) almost completely prevents evaporation.

4. A first inspection will show whether the sulphate of copper is used up or not; the electro-motive force is kept up until the last crystal has disappeared and the resistance rather diminishes.

Lately, Meidinger has replaced the copper with lead. This does not in any way alter the nature of the chemical action, for the electrode soon becomes covered with the deposit of copper. The copper wire which serves as the connection is also replaced by a strip of lead, which indeed needs no protection, for lead is not attacked by the liquids which enter into the composition of the battery.

The only advantage in the substitution of lead for copper is that it reduces the cost of the battery.

We have already spoken of the disadvantage in keeping the solution in a state of saturation; we find a confirmation of our opinion in the following fact: The administration of Berlin recommends not to recharge the cells until the last crystal of the preceding charge has disappeared and not to put in too large a quantity of sulphate of copper.

A commission chosen by this administration observed, also, that the economical coefficient diminished when the cells were too frequently charged or with too great a quantity.

According to our view, there should have been a complete suppression of crystals. We will soon see that the commission arrived at a contrary conclusion; it preferred, it would seem, to sacrifice the economy of the material and to diminish the work of the operators.

Dr. Dehms reports the resistance of Meidinger's element as varying from 4 to 9 of Siemen's units,* according to the models.

MEIDINGER'S FLASK BATTERY.

Instead of the interior tube reservoir, Meidinger has finally adopted a large exterior flask as reservoir for the crystals (Fig. 32). This form has been in use a long time in Bavaria and Germany, where it was recommended for branch telegraph offices and for the following reasons:

1. In all branch offices (in Northern Germany) continuous currents are employed; the consumption of the battery is therefore considerable, and if the labor of frequently renewing the elements is to be spared, a battery should be adopted which contains a large quantity of material. This is found in the battery in question, as the flask holds two pounds of sulphate of copper.

2. The disappearance of the charge of sulphate of copper is here more plainly seen than in the other model, and cannot escape the most rapid glance.

3. In these branch offices the battery is rarely made to

---

* A Siemen's unit is equal to the resistance offered by one metre of mercury whose sectional area is one millimetre.

work simultaneously upon several lines; therefore its resistance, which is considerable (10 of Siemen's units to each cell of the approved form), presents but little inconvenience.

Under these two forms, Meidinger's battery is employed

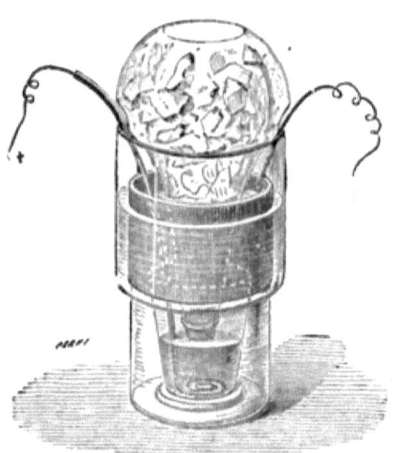

Fig. 32.

in the railway and state telegraph services of Russia, almost to the exclusion of all others.

The batteries are generally left one year without care, in some instances fourteen months. But in offices where there is a very active service, the sulphate of copper is renewed every four or six months.

KRUGER'S BATTERY.

This is a battery said to have advantages over other batteries without the porous jar. The zinc is placed as in Callaud's; but the copper has the form of a hollow cylinder, and is placed vertically in the jar and has the same height. The cylinder is made of a very thin piece of copper, cut longitudinally at the bottom in six places; the divisions thus formed are bent outward and hold the cylinder in the centre of the jar. There are also three pegs fastened in the zinc, which serve to keep the copper cylinder in place.

If sulphate of copper crystals are added, they are placed in the copper cylinder and the solution formed remains at the bottom of the jar.

This disposition costs less than that of Meidinger, and produces a battery with much less resistance; in this respect it is preferable to Callaud's, but costs more.

We know, however, from good authority, that this battery presents a grave inconvenience. The copper becomes eaten at the line of separation of the two liquids, and at the end of a certain time it breaks. It will be remembered that we mentioned, when speaking of Callaud's battery, the necessity of protecting with gutta-percha the copper wire which traversed the liquids and served as the connection.

This inconvenience could be avoided by substituting lead for the copper, as Meidinger sometimes does.

In the first edition of this work we expressed the opinion that the portion of copper which is not immersed in the solution of sulphate of copper tends to increase the internal conductivity of the element, although it takes no part in the regular chemical action which produces the current. Since then, Mr. G. d'Infreville has sent us the very interesting results of his experiments upon this subject. He has found that this disposition causes a diminution in the electro-motive force. This will not seem surprising to physicists, for they will readily understand that in Kruger's or any analogous battery the circuit becomes closed by the sulphate of zinc at the top; a constant derived current is thus produced, and consequently the difference of potential is diminished between the electrodes.

SIR WILLIAM THOMSON'S BATTERY.

This illustrious physicist invented a very original disposition of the Daniell gravity battery. The elements are piled up one upon another as in Volta's column battery or as in Marié Davy's sulphate-of-lead battery.

These elements (Fig. 33) consist of wooden trays, lined

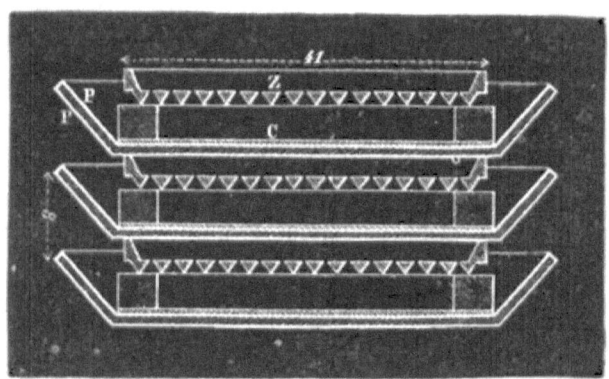

FIG. 33.

on the inside with lead to make them water-tight. At the bottom of each tray is placed a thin plate of copper. In the four corners of this square tray are little blocks of wood which support the zinc electrode. This latter has the singular form of a gridiron, having its bars very close, but still leaving space enough between for the circulation of the liquid. The feet of this gridiron are turned upwards, supporting the cell above.

In some instances the zinc is wrapped in parchment

paper, thus constructing a porous jar which prevents the mingling of the liquids; this may, however, well be dispensed with.

The connection between one cell and the following one is simply obtained by their weight, which presses the lead on the bottom of each tray upon the four corners of the zinc below.

Care must be taken in charging the cells, to place them in a perfectly horizontal position; this can be easily ascertained by pouring some water in the tray and observing if it spreads equally over the bottom.

The sole advantage of this form is the feeble resistance it gives to the cells. It is used as local battery in the submarine telegraphs, where Sir William Thomson's "Siphon Recorder" is at work. We are informed that it is also employed in Russia, and has indeed been applied for electric light purpose.

The battery should be charged with a solution of sulphate of zinc, whose density is 1.10 and the sulphate of copper crystals are placed as regularly as possible all around and at the bottom of the tray.

Scarcely more than eight or ten cells can be piled up in one column, and a series of these columns are connected by very large conductors. If the battery is used in delicate experiments or for the working of the "Siphon Recorder," it must be constantly watched, and a little of the sulphate of zinc solution taken out daily and replaced with pure water. If possible, the density of the liquid should be kept between 1.30 and 1.10; for that purpose it should be measured from time to time with an hydrometer.

The circuit of the battery should always be closed, in order to avoid the deposit of copper upon the zinc. It is advisable when the battery is not in use to keep it in short circuit, so that the sulphate of copper may be rapidly exhausted.

It is perhaps a good idea to measure, now and then, the electro-motive force of each series and of each cell separately. When one cell is found to have lost much, on account of the deposit of copper upon the zinc, it should either be taken away entirely or placed in short circuit, to get rid of the useless resistance it introduces into the circuit.

Sir William Thomson has, besides, given an inverse form to his battery, upon the basis that a saturated solution of sulphate of zinc has a density of 1.44, whereas a saturated solution of sulphate of copper has a maximum density of 1.18; the sulphate of zinc is below and the sulphate of copper is above, the contrary of that which prevails in all preceding batteries. We have not been able to obtain any notes that the eminent author may have published upon the subject, and do not therefore know his motives for this reversed form. We only know that it is practically inconvenient, and presents the following fault, viz., that reduced copper falls upon and finally covers the zinc. It is possible, however, that this reversed form possesses great advantages for special cases.

The principle of this form deserves notice. It is seen that if, in ordinary gravity batteries, the sulphate-of-zinc solution is permitted to approach or to arrive at saturation, the two liquids are no longer separated by their difference in density in the desired manner; the result is that the battery works badly and there is considerable waste.

ELECTRO-MOTIVE FORCE OF THE DANIELL GRAVITY BATTERY.

At first it would seem that the electro-motive force of the Daniell battery ought to be the same, with or without the porous jar. The results of all measurements show, however, the superiority of the gravity battery. One might believe that these differences arise from errors of observation; but we have assured ourselves by a direct comparison made by the method of opposition that the electro-motive force of gravity cells is the same in models differing widely from each other, and that it is much greater than that of a cell with the porous jar.

This is a peculiarity difficult to explain, unless a feeble polarization at the surface of the porous jar be admitted.

It must undoubtedly be placed with that other fact of which we have spoken; viz., the electro-motive force of Daniell's porous-jar element is very inferior to its normal value when it is first mounted and when the imbibition of the porous jar is not yet complete.

CHAPTER III.

GENERAL REMARKS UPON DANIELL BATTERIES.

AMALGAMATION OF ZINC IN THE DANIELL BATTERY.

We have, from the beginning of this work, shown the advantages of amalgamated zinc over commercial zinc when immersed in dilute sulphuric acid.

When, therefore, a Daniell battery is charged with dilute sulphuric acid, there is a great advantage in the use of amalgamated zinc; but to-day the acid is mostly suppressed and replaced by sulphate of zinc. It is generally believed that it is still advisable in this case to amalgamate the zinc, but we have informed ourselves as to the subject, and find that the addition of mercury is rather injurious than useful.

We have taken several cells, one having amalgamated zinc, joined them in intensity, and allowed them all to undergo the same alternatives of rest and action. At the end of fifteen days it was found that the local action had been much greater upon the amalgamated zinc than upon the others; that is, the deposit upon the surface of the zinc and at the bottom of the jar was more abundant in the amalgamated cell.

To confirm this result we have made the following experiment, like that of De La Rive: We put a piece of well-amalgamated zinc in a solution of sulphate of copper; the attack took place immediately and rapidly; in twenty minutes all the sulphate of copper was decomposed.

These experiments possess no other interest, however, than that of justifying the general practice adopted by those who use the Daniell battery without free sulphuric acid; that is, in the electric telegraph and analogous applications.

Experiments of Jules Regnauld give the following figures for the electro-motive forces of the batteries in question:

Pure zinc, sulphate of zinc, sulphate of copper, copper.... 175
Amalgamated zinc, sulphuric monohydrate, 1 vol.
Water, 10 vol.; sulphate of copper, copper................ 179

It is therefore clear that the use of amalgamated zinc in acidulated water occasions a slight increase of the electro-motive force; this superiority, however, which is less than $2\frac{1}{2}$ per cent, has but little practical interest.

COPPER-PLATING.

The study of Daniell's battery has led to the creation of a vastly important industry; namely, that of copper-plating. Its object is the reproduction in copper of artistic or mechanical models, typographic blocks, medals, bas-reliefs, statuettes, etc. etc.

Jacobi in Russia, and Spencer in England, having observed that the copper which is deposited upon the copper electrode was so fine as to reproduce the smallest irregularities of surface of that electrode, decided to make use of this process of moulding, and they have shown its great utility. The extreme facility of the process renders it accessible to every one, and consequently its use has spread with infinite variations.

We shall not enter upon the details of this process, but only show the most simple method of copper-plating, which consists in the use of a large Daniell cell with the porous jar. The porous jar contains the amalgamated zinc immersed in well-acidulated water. In the outside jar, which is comparatively much larger, is a saturated solution of sulphate of copper and electrodes formed of pieces of gutta-percha moulded upon the model (medal or wood-engraving). To the surface of these gutta-percha moulds is imparted a conductive quality by means of an impalpable powder of plumbago spread upon it with brushes. An exterior conductor joins the zinc to the negative electrode. The deposit commences directly and is very slow at first upon the plumbago, but as the copper accumulates the process advances at a more rapid rate.

All the figures which illustrate this book were first engraved on wood, then moulded in gutta-percha, and finally reproduced in copper, in the manner we have just described. It is with these *electrotypes* that the impressions were made, and it is the process generally employed.

It is seen that the battery which produces this industrial deposit is one of a single cell, and that the deposit takes place in this single cell.

It is impossible to imagine any more simple process; but it is not as economical as it might at first appear, because, to extract a given weight of copper from the sulphate, and deposit it upon the mould, there must be consumed an equivalent weight of zinc and sulphuric acid, without counting the mercury lost in the manipulations.

IRREGULARITY OF THE CHEMICAL ACTION IN DANIELL'S BATTERIES.

We said in the beginning that the chemical action in the Daniell battery was limited to the dissolving of the zinc, the substitution of zinc for the copper in the sulphate, and the deposit of copper upon the conducting electrode. That is indeed the theoretical and principal action; but it is not the only one. We have shown in the foregoing that there are local actions in the neighborhood of the zinc and a deposit of copper upon the zinc.

But that is not yet all. If the Daniell battery be closely examined (Callaud's battery is well suited to this examination, because one can see everything that goes on inside), a continual formation of gaseous bubbles at the surface of the zinc, and indeed upon the copper electrode, is seen.

As to the freeing of hydrogen from the zinc, the following is the simple explanation :

This zinc is immersed in sulphate of zinc. The zinc not being pure, small voltaic cells form themselves at its surface, and consequently the water is decomposed and hydrogen given off.

The explanation is analogous concerning the copper electrode; the different parts of this electrode are immersed in unequally saturated portions of the sulphate of copper, or even in portions containing sulphate of zinc ; a voltaic cell is thus constituted between the two liquids and the single electrode. We will return, at the end of this work, to identical electrode batteries, when all doubts the reader may here have will be dispelled.

There is less freeing of hydrogen from the copper than

from the zinc, no doubt because the voltaic cells formed have a much smaller electro-motive force.

Whatever the truth of these explanations may be, the fact of the freeing of hydrogen is not to be doubted. Large gaseous bubbles are seen attached to the surface of the zinc, and in perfect stillness they may be heard to free themselves now and then and rise to the surface, making a slight noise as they explode.

These abnormal actions are not the only ones that take place in the Daniell, but they are the most important.

CHAPTER IV.

BATTERIES DERIVED FROM THE DANIELL.

By replacing the copper in the Daniell battery by some other metal and the sulphate of copper by the sulphate of that metal, batteries possessing qualities analogous to those of the Daniell may be made.

Several of them have a practical interest, but we will at first mention two that can only serve in laboratories.

The cadmium battery is formed of zinc, sulphate of zinc, sulphate of cadmium, and cadmium. Its electromotive force is 0.31; that is, thirty-one hundredths of the unit, or, to use round numbers, one third of the Daniell battery. Such are, at least, the figures given by M. Jules Regnauld.

The aluminium battery is still more feeble: zinc, sulphate of zinc, aluminic sulphate, and aluminium. The electro-motive force is 0.2, or one fifth of that of the Daniell.

The tables at the end of this work will show other combinations, which are of no interest here.

To our knowledge, no one has ever tried the zinc, sulphate of zinc, sulphate of iron, and iron battery; a study of it would undoubtedly be very interesting.

The batteries which we will now study have indeed the Daniell for model, but they possess a striking peculiarity; viz., the depolarizing salt is almost insoluble, which gives rise to certain interesting results.

MARIÉ DAVY'S SULPHATE-OF-MERCURY BATTERY.

Of all batteries having the Daniell for model, the most interesting is that proposed by Marié Davy. It has been extensively employed, and is still used in many cases.

Let us substitute, in the Daniell, sulphate of mercury for sulphate of copper, and carbon for copper, and we will have the new battery. In truth, if we had strictly followed the model, we should have replaced the copper by mercury; but the liquid nature of this metal and its high price have caused the preference of a carbon electrode. Besides, the action of the battery producing the reduction of the sulphate and the deposit of metallic mercury upon the negative or conducting electrode, the difference disappears in time, and there is in reality an electrode of mercury in which is immersed a piece of carbon. It has been proved that the electro-motive force is not changed by the suppression of the carbon and the use of mercury, provided the latter is pure.

Fig. 34 represents the form that Marié Davy has given to his cell. The zinc is a hollow cylinder placed in the glass jar around the porous jar; this latter contains the carbon electrode surrounded with a liquid paste of sulphate of mercury. The carbon is capped with copper, to which is soldered a strip of the same metal, connected with the zinc of the adjoining cell. The carbon may also be capped with lead, and the connection made in the same way. In either case it is better to first dip the top of the carbon in a bath of paraffin, in order to fill up all the pores so that no liquid may rise by capillarity and attack the lead or copper.

The chemical action in this battery is precisely analogous to that in the Daniell. The sulphate of mercury is reduced, an equivalent quantity of sulphate of zinc is formed, the zinc is dissolved, and metallic mercury is deposited in the porous jar, either upon the surface of the carbon-electrode or in the mass of the sulphate of mercury.

In the French Telegraph, suboxide of mercury

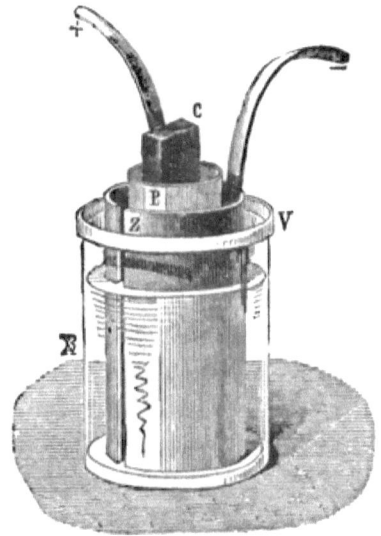

Fig. 34.

($SO_3 Hg_2 O$) is generally used; consequently two equivalents of mercury are set free for one of zinc dissolved.

But there are other sulphates that may also be employed.

The sulphate of protoxide (HgO, SO_3) is frequently used for medical purposes. This salt presents a singular peculiarity: coming in contact with water, it decomposes into two salts, the one basic and but very little soluble;

the other acid, very soluble, which has not as yet, to our knowledge, been analyzed. The first salt has a yellow appearance, whose formula is $3(HgO) SO_3$; it is therefore a tribasic salt.

Marié Davy's batteries may be charged with monobasic sulphate of protoxide ($HgOSO_3$), and even with the tribasic sulphate alone [$3 (HgO) SO_3$], and we are assured that these two batteries have sensibly the same electro-motive force as that in which sulphate of suboxide (Hg_2OSO_3) is used.

It is said that this last salt ought to be preferred to the others, because the tribasic salt, mentioned above, breaks the porous jars by its little solubility; but admitting this, the two salts of protoxide can very well be used in gravity batteries, of which we shall speak further on.

Marié Davy's battery presents the following advantages:

1. It has an electro-motive force very much superior to that of Daniell; it is represented by 1.5.

2. The slight solubility of the sulphate of mercury causes but a very slow diffusion in the outer liquid; the result is that the local actions and the waste are not so important as in the Daniell battery.

3. The mercury which is deposited upon the zinc by local action, and without any corresponding production of electricity, amalgamates the zinc, or keeps up its amalgamation, which is very useful and prevents the waste of the sulphuric acid if it is used in the outside jar.

The faults of this battery are as follows:

The sulphate of mercury is a violent poison; the price of this salt is high and very variable, as the price of the mercury itself. Finally, it is apt to weaken under certain circumstances, as we will show in detail.

WEAKENING OF THE SULPHATE-OF-MERCURY BATTERY.

When Marié Davy's battery is employed for the telegraph or any intermittent service, no variation in the electro-motive force is noticed, and the battery may be regarded as constant. But if the circuit remain continuously closed, or if the intervals of its being open are too short, a diminution in the intensity is observed. This diminution is the result of several causes. We will commence by speaking of one of them, which we have not yet met in the course of this study.

The special characteristic of the Marié Davy battery, when compared with the Daniell, is the insolubility of the mercurial salt. It must, however, be noted that the sulphate of mercury is not altogether insoluble, but very little soluble, and it would be a great mistake to call the battery which we are now studying a single liquid battery. There are, in reality, two liquids, the solution of sulphate of zinc and that of sulphate of mercury. In 1860 we made the following experiment:

A solution of sulphate of mercury was prepared and well filtered, so that no traces of undissolved salt might remain; then a battery charged with zinc, sulphate of zinc dissolved, sulphate of mercury dissolved, and carbon. This battery had exactly the same electro-motive force as that of an ordinary Marié Davy battery containing 100 grammes (or more) of paste of mercurial salt.

The comparison was made by opposition with a very sensitive galvanometer; there was, therefore, no room for any doubt.

But if this filtered solution of sulphate of mercury

battery were called upon to furnish a current, it would be seen to weaken very rapidly, which is easily understood.

The very small quantity of mercurial salt dissolved was soon exhausted, and depolarization was no longer effected; from that moment it was simply a single-liquid battery which polarized promptly.

That is, in reality, what may be expected in all batteries containing an only slightly soluble depolarizing salt, when they furnish a large quantity of electricity. If the consumption of the dissolved mercurial salt is more rapid than the dissolving of the salt in the liquid, the solution of course weakens, and finally does not act at all, so that the battery is no longer a two-liquid battery but a single-liquid one, and consequently becomes weakened by polarization.

The above constitutes the first period of the action of the battery. We will now examine the second. As soon as the battery is reduced to a single liquid it becomes subject to polarization. This polarization is subject to the rules that we have indicated in the first part of this work. It depends upon the intensity of the current, and assumes a much greater value when the cell in question is polarized by the current furnished by other cells joined with it in intensity. It depends upon the dimensions of the cell and of the elements. And it depends, finally, upon the length of time of the experiment and upon the resistance of the circuit.

This is the same phenomenon that we have already studied several times. But a third period has been observed in this battery, and also in others. When a cell becomes extremely polarized it loses all its force, and then electrolysis of the sulphate of zinc takes place. This salt is decomposed, zinc is deposited upon the carbon, and, coming in contact with the mercury there, the two unite and form one amalgam of zinc, which is at first positive, but which soon becomes negative, as compared with the zinc, when the quantity dissolved in the mercury becomes considerable. Thus is formed a zinc amalgam-of-zinc cell whose poles are contrary to those of the original cell. In other words, the poles are reversed in this third period. We have previously pointed out an analogous action as taking place in the salt-water battery, and it is probable that the cause is the same. Now this reversing of the poles and the extreme polarizations do not take place in the general operations of the telegraph, because the cells are only used intermittingly and have time, during the intervals of repose, to depolarize.

It is understood that the smaller the cells and the less mercurial salt they contain, the more rapid the weakening of the solution of sulphate of mercury. Therefore cells $4\frac{3}{4}$ inches high are recommended. Cells of these dimensions, however, cost a great deal, and in general cells $3\frac{3}{20}$ inches high are adopted, which are all that is necessary for unimportant telegraph offices.

Care should be taken that the level of the liquids remain sensibly the same in cells joined in intensity, because if in one the level is notably lowered, it is as if this cell had become smaller; the dissolved mercurial salt of this cell would become rapidly exhausted and the cell would polarize.

SULPHATE-OF-MERCURY GRAVITY BATTERY.

A French physicist arranged sulphate-of-mercury cells after the model of Callaud's battery, and succeeded very well by adding to the mass of mercurial salt fragments of crushed gas-retort carbon (volume for volume). This car-

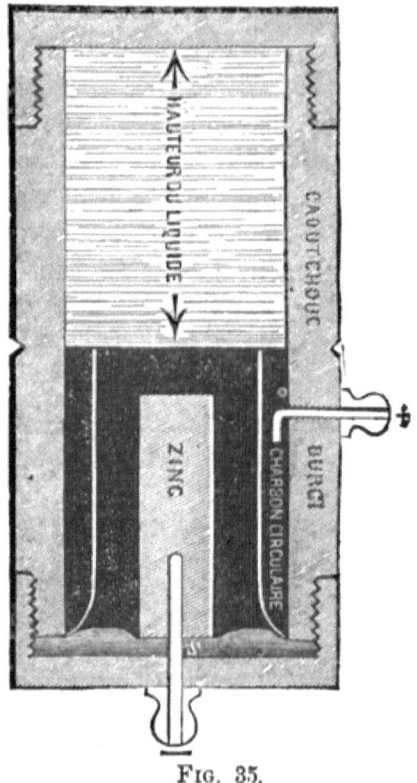

Fig. 35.

bon produces a kind of drainage and prevents the salt from becoming too compact, which in ordinary batteries renders the dissolution slower than is proper.

We believe that the tribasic salt of sulphate of protoxide of mercury, which is less costly than the other mercurial sulphates, might well be employed in batteries of this kind.

TROUVÉ'S REVERSIBLE BATTERY.

Trouvé arranged, with a view to medical purposes, a battery which presents incontestable advantages for certain uses.

A cell of this battery is shown in Fig. 35. The outside jar is a cylinder of ebonite, closed at both extremities by ebonite screw tops. At the upper part is seen the zinc in the shape of a small cylinder fixed in the middle, to which is attached a wire that passes through the top and constitutes the negative connection of the cell. The carbon is cylindrical and surrounds the zinc. The liquid, formed of water and of sulphate of protoxide of mercury (SO_3HgO), does not reach the lower part of the zinc, when the cell is placed, as shown in the figure; but if it be turned upside down or put upon its side, the liquid comes in contact with the electrodes and the current begins to flow.

The cell is hermetically closed, and there is no danger of any leakage. It is very convenient for many purposes, and forms part of some Volta-induction apparatus made by Trouvé.

GAIFFE'S BATTERY.

Gaiffe uses the sulphate of mercury battery for his Voltaic-induction apparatus. Two cells are joined, as Fig. 36 shows. Each cell is contained in a small separate vessel of ebonite, at the bottom of which is a carbon plate. Upon the carbon is placed some water and sulphate of protoxide of mercury. A small plate of amalgamated zinc is immersed in this liquid, and is furnished with a little knob in the centre by which it may be lifted out. The zinc rests upon platinum wires fastened in the

Fig. 36.

moulding of the ebonite, which establish the communication with the carbon of the adjoining cell.

These batteries may work for about an hour, when the sulphate should be removed. This liquid is generally freshly made every time the induction apparatus is used, and the old water and yellow sulphate are thrown away.

LATIMER CLARK'S STANDARD BATTERY.

This eminent electrician proposed to "discover a form of the Voltaic battery having a perfectly constant electro-motive force and maintaining an invariable difference of potential between its poles."*

He has found that the voltaic combination which best fills these conditions is an element composed of zinc, sulphate of zinc, sulphate of mercury, and mercury.

The zinc ought to be chemically pure: distilled zinc is used.

The formula for the sulphate of mercury is Hg_2OSO_3; it is a white salt that does not become yellow by the ad-

---

* Philosophical Transactions of the Royal Society, June 19, 1873.

dition of water. It is prepared by dissolving pure mercury in warm sulphuric acid, but not boiling; it must be carefully washed, because the presence of any free acid would notably change the results. The salt must contain no sulphate $HgOSO_3$ (sulphate of protoxide), recognized by its transformation into a yellow salt by the action of water.

The sulphate of zinc should be pure and used in a state of saturation; it can be easily understood that its composition can thus be rendered more constant.

The battery is prepared as follows: The sulphate of zinc is dissolved in distilled boiling water, left to cool, gently poured off, and the saturated liquor thus obtained is used to form a thick paste with sulphate of mercury; this paste is heated to 100° centig. in order to expel any air it may contain. It is then poured upon the previously heated surface of the mercury, the zinc is suspended in the paste, and finally the jar is closed with melted paraffin.

The positive connection is a platinum wire passing through a glass tube and descending to the mercury. A better plan is to place this positive connection in an outside glass tube which opens into the jar near the bottom.

"The electro-motive force of these elements is remarkably constant, provided they remain open and be not weakened by work. Numerous experiments have been made between new ones and others which had been used several months, and it was found that the greatest differences did not reach one thousandth part of the total value of the force."

Experiments have been made to determine the variations in the electro-motive force with the temperature. The average showed a diminution for an increase of temperature in the proportion of six hundredths per degree centigrade. The variations are more marked as the thermometer nears 0° centig., about eight hundredths per degree; from 5° to 25° they are about .06 per cent, and about .055 per cent up to 100° centig.

Mr. Clark has determined the electro-motive force in absolute measurement, and has found it to be 1.4573 volts, and 1.4562 volts with a sine galvanometer, the temperature being 15.5° centig.

SULPHATE-OF-LEAD BATTERY.

This is the same kind of battery as the sulphate-of-mercury battery. It is a Daniell with an almost insoluble depolarizing salt; it is formed of zinc, sulphate of zinc, sulphate of lead, and lead.

The cheapness of sulphate of lead has caused this battery to be extensively used, but it has finally been abandoned for certain reasons, which we will give farther on.

M. Becquerel was the first to try this combination. His conducting electrode was either lead or a piece of carbon, a piece of copper or tin.

In 1860 Marié Davy proposed a new form which has been abandoned; it deserves notice, however, as it was a good imitation of Volta's column battery, and has been imitated in turn by Sir William Thomson in his large Daniell gravity battery.

Marié Davy's elements consisted of pans of tinned iron provided with three arms placed horizontally and at equal distances apart. Upon the under side of each pan was soldered a disc of zinc covering the whole bottom. In each pan was a layer of sulphate of lead about $\frac{1}{8}$ of an inch thick and a portion of pure water or water containing a little sulphate of zinc. The pans are placed one above the other, so that the zinc of one is immersed in the liquid of the other below it. It is necessary to keep the successive cells at equal distances apart, which is done by means of vertical wooden supports to which the horizontal arms are fastened. A battery of forty cells of this kind forms a column about 40 inches high. This form, however, has not been preserved; it weakened rapidly, and we believe that its principal defect lay in the too small quantity of water it contained.

M. Edmond Becquerel gave to the sulphate-of-lead battery the regular form of porous jar-cells. He arranged a solid cylindrical mass of sulphate of lead around a central piece of lead from $\frac{1}{8}$ to $\frac{1}{4}$ of an inch in diameter. Consistency may be given to this mass by mixing it with chloride of sodium (100 grammes of dried and crushed sulphate of lead and 20 grammes of sea-salt), and by wetting it with a saturated solution of sea-salt (50 cubic centimetres). This cylindrical mass of sulphate of lead may be surrounded by a layer of plaster, which serves as a porous jar.

Finally, ordinary porous jars may be used containing the lead and the sulphate.

One of the faults in this battery is that its electromotive force is about one half that of a Daniell, while its resistance is very great. The result is that, in order to obtain the same intensity, a number of cells more than double that of the Daniell battery is required. It would seem, from the foregoing, that this battery would be badly adapted to an electric-bell service, as the bells are placed in circuits of little resistance, and their electromagnets themselves have but little resistance; therefore the internal resistance of the battery is greater than that of the outside circuit, which is a very unfavorable condition. It is, however, to this use that the battery in question has been put, and it must be that it does all that is required of it.

It would seem that the economy resulting from the use of a cheap salt would be more than balanced by the necessary number of cells, and by the cost of so much zinc and so many glass and porous jars. This consideration must have finally attracted attention; for sulphate-of-lead batteries have been gradually abandoned.

WEAKENING OF THE SULPHATE OF LEAD BATTERY.

There happens in the battery in question precisely what we have seen in Marié Davy's battery: the sulphate of lead does not act unless dissolved. This dissolution is slow, and if the consumption of electricity is active, and consequently that of the dissolved sulphate of lead, the solution weakens and the battery ceases to be depolarized; in other words, the battery becomes weakened.

It is probable that if an energetic current were made to act upon the battery thus polarized, the sulphate of zinc would be electrolyzed and zinc would be deposited upon the lead, as in Marié Davy's cell. This observation has no practical interest, but only tends to show once more the great similarity of the sulphate-of-lead battery to the mercurial-salt battery.

If the sulphate-of-lead battery were again to be used, we believe that there would be an advantage in putting pieces of gas-retort carbon in the mass of sulphate of lead; this would facilitate the dissolving of that salt which is the condition of depolarization, and it would also give a greater conductivity to the mass which fills the porous jar.

VARIOUS SALT BATTERIES.

By substituting for the sulphate of copper and zinc in the Daniell the corresponding nitrates, or the acetates, or the chlorides, simple modifications of the Daniell are obtained. None of these batteries have any practical interest, on account of the high price of the materials.

M. Jules Regnauld has carefully measured the electro-motive forces of several of these batteries, and the comparison of the figures is worthy of attention.

Pure Zinc	Copper.	Electro-motive force expressed in Volts.
Sulphate of zinc	Sulphate of copper	0.955
Nitrate " "	Nitrate " "	0.873
Acetate " "	Acetate " "	0.955
Chloride " "	Chloride " "	0.955

The equality of these figures is very interesting. But from these particular instances no law of physics can be deduced: other measurements of M. Regnauld show that this equality does not exist in all analogous series of batteries.

CHAPTER V.
ACID BATTERIES.

IN the beginning of Part II. of this work we indicated how the electrode may be depolarized by means of substances rich in oxygen and easily decomposed, notably highly oxygenated acids. The experiment which we then cited is due to Grove, and has led to the production of one of the best batteries known.

GROVE'S BATTERY.

To remain faithful to our method of exposition, we should consider Grove's battery as one of the derivatives of Volta's. In the latter the zinc is attacked by the dilute suphuric acid, and hydrogen is evolved upon the conducting electrode, copper, platinum, or carbon. If this platinum electrode be surrounded by nitric acid, the latter is decomposed, oxygen is set free and forms water with the polarizing hydrogen, and nitric oxide is given off. The battery thus modified is without polarization, or, in other words, is constant. It is known by the name of Grove's battery, and dates from the year 1839.

That which is commonly called Grove's battery contains a conducting electrode of platinum; it is understood, however, that Grove's conception is a general one and may be easily modified. Fig. 37 represents this battery under its most common form.

A square porcelain jar is used as the outside recipient;

it contains a well-amalgamated zinc electrode, having the shape of an U. Inside of the zinc thus shaped is placed a porous jar containing a very thin piece of platinum, which serves as the conducting electrode; water acidulated with sulphuric acid is put with the zinc in the outside jar, and nitric acid fuming is poured in the porous jar. The platinum is connected, as shown in the figure, with the projecting portion of the zinc in the adjoining cell.

Such is the form of Grove's cell employed in England

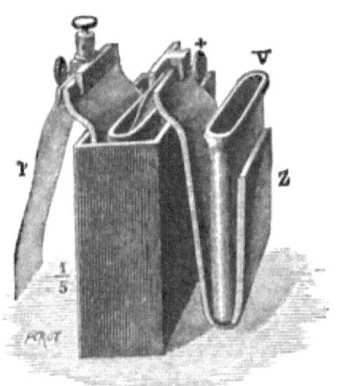

FIG. 37.

to the entire exclusion of Bunsen's form, so generally used in France. We will speak of the latter farther on.

In Germany, Poggendorf's arrangement of Grove's cell is used. The porous jar is cylindrical and contains the platinum in the shape of an S, in order to offer as large a surface as possible. The platinum is fastened to a porcelain stopper which almost completely closes the porous jar. This arrangement is represented by Figs. 38 and 38a. We will not dwell any longer upon it, because we believe that Grove's battery has received but a limited application in Germany, and that Bunsen's disposition is generally preferred.

Our main object is more to make known those batteries which, in one country or another, are the most exten-

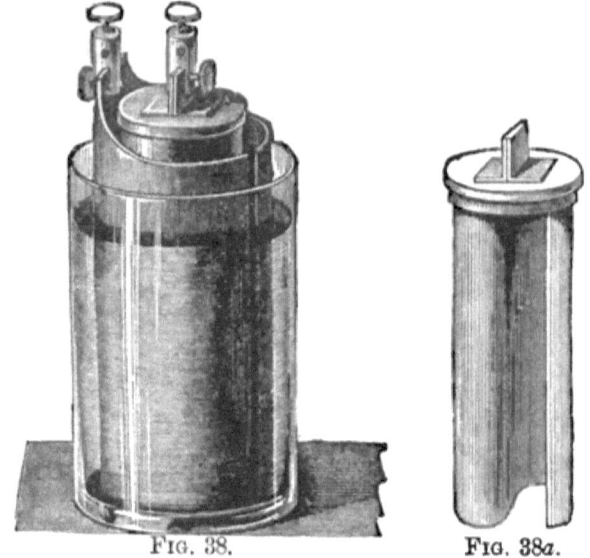

Fig. 38.   Fig. 38a.

sively employed, and not so much to treat of the many combinations which are only theoretically interesting.

CHEMICAL ACTIONS IN GROVE'S BATTERY.

We have said that the zinc becomes oxidized at the expense of the water and forms sulphate of zinc; that the hydrogen of the water thus decomposed reduces nitric anhydride (N_2O_4) and produces nitric oxide (NO). This latter gas, coming in contact with the air, is transformed into nitric tetroxide, recognizable by its color and by its suffocating properties. It is absolutely certain that the Grove batteries produce nitric tetroxide; therefore the decomposition of nitric acid must surely produce nitric oxide, as we said above; but that this is the only form of decomposition that takes place is not certain. On the contrary, it is probable that nitric trioxide is produced. This appears to be the result of theoretical considerations, into the details of which we cannot enter.*

Other actions take place in Grove's cell, one of which is the formation of ammonia. If, in fact, after the liquids are exhausted they be evaporated, it is observed that the addition of lime in the concentrated liquid produces an abundant freeing of ammoniacal gas. This proves that if a portion of the hydrogen has combined with the oxygen of the nitric acid to form water, another portion has combined with the nitrogen to form ammonia.†

Unfortunately batteries have not been completely examined from a chemical point of view, and it is not known precisely what takes place within them. To completely elucidate the question, the gases evolved in each cell should be separately collected; then these gases and also the matter left in the liquids should be analyzed.

PRACTICAL DETAILS.

The platinum electrodes are generally 5 in. high by 2 in. wide, and they must not be too thin, for the resistance would thus be increased in a manner extremely detrimental to the desired results.

There is no serious disadvantage in permitting the porous jar to touch the zinc, and on the other hand it is very important to diminish as much as possible the dis-

---

* See Daniell, "Introduction to Chemical Philosophy."
† See Gavarret, "Traité d'Electricité," t. ii. p. 446.

tance between the electrodes and consequently the resistance of the cells.

The U form of the zinc is not very economical, because it breaks at the bottom before it is worn at the top. It has been proposed to place some mercury in the bottom of the porcelain jar and to use two zinc plates, one on each side of the porous jar, which are thus united by the mercury at the bottom. This disposition, which we have already indicated for other batteries, allows the more complete consumption of the zinc.

BUNSEN'S BATTERY.

In the beginning, Grove thought of substituting charcoal or even gas-retort carbon for the platinum, and several public experiments were made in London; these, however, had been forgotten when, in 1843, Bunsen became possessed of the same idea and succeeded in introducing the general use of his arrangement.

Without further insisting upon the history of this invention, we will first describe Bunsen's battery as it is used in France, and then the form employed in Germany, which is very like that originally proposed by Bunsen.

FRENCH MODEL (Fig. 39.)

The outside jar is made of glazed earthenware, which is less fragile than glass, and this consideration is important in the use of a battery which is often moved from one place to another at greater or less distances. For batteries which always remain in the same place there is no inconvenience in the use of glass jars; but as their cost is very little less than that of the glazed earthenware

ACID BATTERIES.

jars, the latter are nearly always used, at least in France. Earthenware subject to fracture should be avoided, for the acid enters into the cracks of the enamel, rendering the jar permeable and fragile.

The zinc is formed of a plate of zinc $\frac{3}{20}$ of an inch thick, rolled in the shape of a cylinder and well amalgamated, which latter precaution allows the zinc to be left

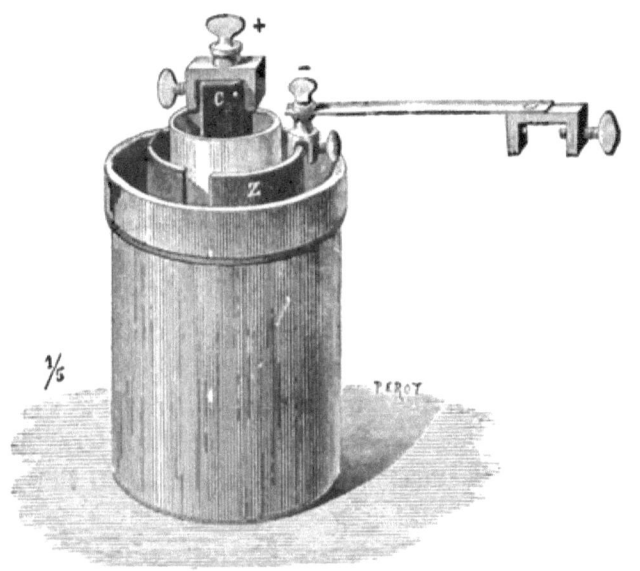

Fig. 39.

twenty-four hours in acidulated water without its being sensibly dissolved.

Some persons solder to the zinc a strip of copper which constitutes the negative connection of the cell; but the two must be riveted together by means of a copper rivet, otherwise the mercury will finally detach one from the other. These means of connection are, however, not to be recommended. The zinc should be higher than the earthenware jar and should be furnished with a binding screw, to which the positive connection of the adjoining cell is attached. This arrangement has the following advantages:

1. Every time the battery is charged, that part of the binding screw which comes in contact with the zinc (and that part alone) should be cleaned with emery-paper. This is very quickly done, and assures to the operator a good contact at those points through which the current passes.

2. When, after having used the battery a certain number of times, the zinc becomes worn at the bottom, it may be turned upside down. Thus the zinc is used up more completely and regularly before it becomes necessary to renew it.

3. The zinc, thus reduced to a simple cylinder having no piece of any kind attached to it is easily packed and requires but a small box for transportation.

The porous jar is placed in the centre of the zinc cylinder and has about the same height. It should be very porous and permit an easy communication of the liquids it separates.

The negative electrode has the shape of a prism with a rectangular base; it should reach above the porous jar, in order to facilitate the connection. We have explained the advantages of the binding screw, and we here repeat that every time the battery is recharged the parts of the binding screw which touch the carbon should be cleaned with emery-paper, in order to secure a perfect contact.

Before the invention of the binding screws, various more or less imperfect means were employed for the connection with the carbon. A conical copper cork was forced into a hole made in the top of the carbon; or copper was deposited upon the head of the carbon, to which a strip of copper was then soldered. Other means, some of which we have already pointed out, were also employed.

All of them presented inconveniences more or less marked, and we do not hesitate to say that they should all be discarded, and the binding screws exclusively adopted. These latter have been used for many years by those persons who successfully produce electric light by means of batteries. We insist upon this recommendation, because a single imperfect contact suffices to cause a notable loss of energy in a battery of fifty or sixty cells.

AMALGAMATION OF THE ZINC.

We have said above that it is necessary to amalgamate the zinc. If this precaution be neglected when the zincs are used with concentrated acids, gases are evolved charged with acid vapors, which render the care of the battery troublesome and which are injurious to the health.

We have also said that by amalgamation the zinc is made more electro-positive, and that consequently the electro-motive force of the cells is increased (an inspection of the tables at the end of this work will make this very plain).

It is therefore necessary that this operation be made with great care, and the simple and regular form of the zinc described above renders the process comparatively easy.

In order to thoroughly amalgamate the zinc, it should first be well scraped and cleaned. The following is the process which we recommend:

The zincs are placed on end in a bucket of water

containing one tenth of sulphuric acid. They stand out of the liquid about half an inch, so that they may be lifted out without immersing one's fingers in the acid. Three zincs are placed at one time in the bucket, in order that each one may remain in the cleansing solution the length of time required for the amalgamation of the other two. A rotation is established in the following manner: Every time that one zinc is taken out to be amalgamated it is replaced by a new one, so that there may be always three in the cleansing solution. The other two have in the mean time been turned upside down, in order to immerse the parts which were exposed to the air.

The vessel containing the mercury for amalgamation has the shape of a portion of a cylinder; it is a little longer than the zincs to be amalgamated. This is the most rational form, for it permits the use of the smallest quantity of mercury. The zincs are carefully immersed in the mercury, the longitudinal opening downwards, in order that the mercury may the more easily reach the interior of the cylinder. The zincs are then turned slowly once or twice, to insure the amalgamation of the entire surface. At the moment of taking the zinc out of the mercury it should be held at an angle of ten or twelve degrees to allow the superfluous mercury to run off. It is then lifted out in a horizontal position with the longitudinal opening upwards, so that no drops of mercury may run off at the angles and thus be lost.

The zincs are finally placed in an empty tub or trough capable of containing several of them; after a certain time a quantity of mercury is found at the bottom, having run off from each zinc.

The vessel used for amalgamation is of enamelled cast-iron, thus being most solid and unalterable.

The zincs ought to be amalgamated but a few hours before the battery is charged, and they ought to be re-amalgamated every time the battery is used, even if there be only twelve hours' interval.

The cleaning requires a little longer time when the zinc is new, and above all when it is covered with layers of adhering salts which remain from preceding operations.

TO MOUNT THE BATTERY.

This operation requires a good deal of method for a battery of fifty or sixty cells, such as is used for electric-light purposes.

The earthenware jars should be placed at short distances from each other, so that they do not touch; they should be placed in one row or in two, or, if room permits, in a circle. The object of these arrangements is to facilitate the filling and emptying of the cells. Place the cells, if possible, upon a table covered with squares of porcelain as are frequently found in laboratories, or, still better, upon rods of glass. In the absence of the above conveniences, the jars may be ranged upon planks of dry wood, but if possible avoid placing them upon the ground in the open air. It can be easily understood that there is an advantage in suppressing all losses or irregular communications between the cells, such as are occasioned if the jars are damp, or if they are placed upon damp earth, or if they touch each other. A striking proof of the existence and importance of these losses is shown in the following phenomenon, which takes place every time the above precautions for insulating are neglected: Whenever one touches any one of the poles of the battery, a shock is felt which is caused by the currents passing into the earth.

The cause of these communications is the formation (upon the surfaces of the jars and supports) of steam and acid vapors, which are abundantly freed from the cells as much by their high temperature as by the chemical action.

As soon as the earthenware jars are arranged, the zincs, the porous jars, and the carbons are put in, the latter being first furnished with their respective binding screws. The connections are then made between the cells, so as to form a kind of chain, commencing with the carbon (positive pole of the battery) and ending with the zinc (negative pole of the battery).

Whether the cells be arranged in rows or in a circle, the screws and connections should be placed so as to be in the operator's way as little as possible when he is pouring in the liquids.

The battery may be thus mounted, without any inconvenience, several hours before using it, since it contains no liquids.

TO CHARGE THE BATTERY.

If the porous jars are new there is an advantage in charging the cells one or two hours before using the battery, so that the liquids may have time to penetrate the porous jars.

If, on the other hand, the porous jars have already served, they have retained liquid acids in their pores, and it suffices to charge the battery half an hour or even quarter of an hour before working it, especially if the battery is to work more than three or four hours.

It is important that the cells be charged in the shortest time possible, in order that those first charged may not be in notably different conditions from the last ones; therefore preparations should be made in advance, and the quickest means employed for pouring in the liquids. Acidulated water should be prepared in a large tub, in order that all the jars may contain the same liquid. Thirty-three jars of pure water (the size of those used in the battery) and three jars of sulphuric acid, at 66° centigrade, are poured into the tub. The water is put in first, and then the sulphuric acid is slowly added, being stirred all the time, in order to render the mixture as homogeneous as possible. This mixture becomes greatly heated, as is known, and if the temperature becomes too high the pouring of the acid should be stopped, and the liquid in the tub agitated with a stick of wood or a rod of glass, or even with one of the amalgamated zincs of the battery.

This mixture can very well be prepared several hours in advance; there is no inconvenience occasioned, however, in using it before it cools off—there is rather an advantage. The only important point is that it be perfectly homogeneous, or, in other words, well mixed.

The liquid is generally drawn from the tub in a pitcher and poured into a funnel held in the hand over the jars; this is, however, very fatiguing for the operator, and necessitates strict attention, in order that the liquid may be at the same level in all the jars. We recommend, therefore, the use of a rubber siphon, like that used in charging the Callaud battery. At one end of the rubber tube is a piece of glass or ebonite, flattened to facilitate its insertion in the jars between the zinc and porous jar. At the other end there is an ebonite mouthpiece, furnished with lead, in order that it may descend to the bottom of the reservoir containing the liquids used to fill the jars.

In order that it may not run too slowly, the reservoir should be placed a metre higher than the jars to be filled. The end of the tube is held in the hand, and it is only necessary to press it with the fingers in order to stop the flowing. This system dispenses with a spigot, produces an instantaneous stop, and allows the regulation, within a fraction of an inch, of the level of the liquids in the cells.

The liquid is started in the siphon in the following manner:

In the hand are held the two extremities of the tube, which hangs below. Water is poured in until it appears at both extremities, then the mouthpiece is quickly placed in the reservoir of acidulated water and a certain quantity of liquid is allowed to run off, in order to purge the tube of the pure water it contains. From that time on everything is ready for the filling of the jars.

This operation terminated, the tube should be well rinsed with water containing a little ammonia, by which precaution it may be made to serve a long time.

For the pouring of the nitric acid a funnel or bottle or any vessel having a spout may be used.

The siphon may also be used, but we recommend Wigner's manner of starting the nitric acid in the siphon, which is done as follows:

The quantity of liquid being smaller, it may be put in a flask whose stopper is fitted with two tubes, one going to the bottom of the liquid and the other not quite reaching to the level of the liquid. By blowing in this second tube the liquid is forced into the siphon, by means of which one person may charge fifty or sixty cells in twenty or twenty-five minutes.

A very neat invention of Mr. Lufbery, for the emptying of casks, flasks, etc., etc. is shown in Fig. 39a.

ACID BATTERIES.

The special stopper, represented apart to the right in the cut, is conical and hollow, and can be adjusted in the mouth of any bottle or flask. This stopper is fitted with an emptying tube, A, and a tube, B; by blowing in the

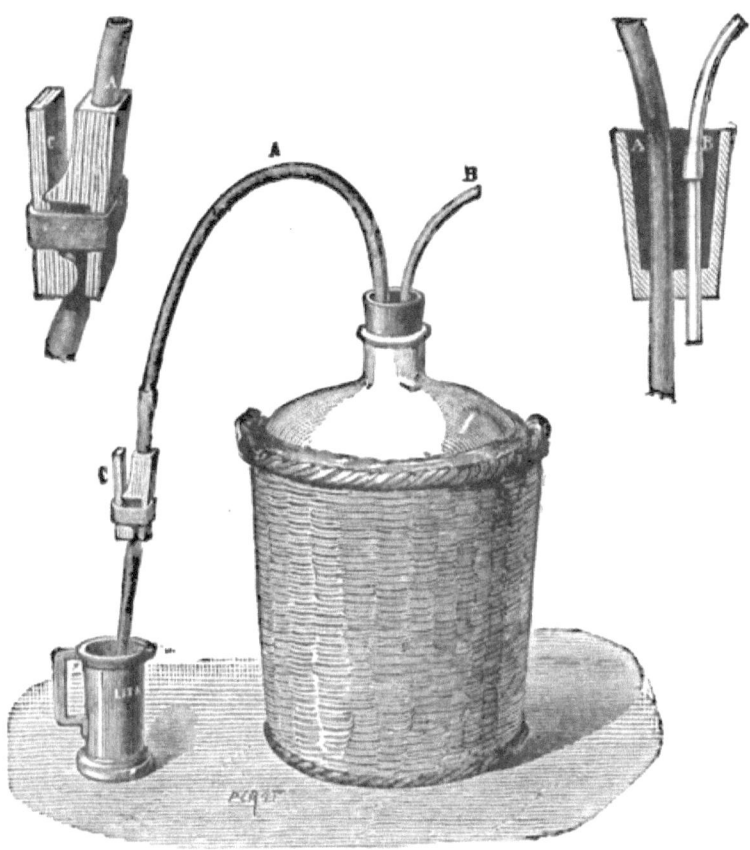

FIG. 39a.

latter the liquid is started in the former. The spigot at the end of the tube A, represented to the left in the cut, is kept closed by a rubber band. It is opened by pressing the upper part between the fingers.

Analogous arrangements are employed in all laboratories for the charging of batteries, and are considered indispensable.

TO DISMOUNT THE BATTERY.

It is plain that as soon as the battery is no longer used, there is an advantage in stopping the waste of the zincs and the acids, which is done by taking the battery to pieces.

It is necessary to have a well-determined method for this operation, which frequently has to be done late at night and with a very poor light. Everything must be done in advance. The zincs and the binding screws are placed in a tub of clear water; if there is time the zincs may be taken out of the water to drip, but there is no disadvantage in allowing them to pass the night in the water.

The carbons must then be taken out and carefully ranged in special earthenware vessels. It is better not to put them in water at first, but to allow them to imbibe acids, which will improve them for the future.

The contents of the porous jars should then be emptied, by means of a funnel, into the vessel destined to contain the acid. This is the most difficult part of the work because of the very abundant vapors given off, which occasion a violent cough if proper precautions are not taken for keeping at a distance.

It is almost impossible to preserve and again use the water acidulated with sulphuric acid, and it may remain all night or longer in the jars without causing any inconvenience.

The nitric acid may be again used if the battery has

ACID BATTERIES.

not worked more than about three hours. By mixing it with some fresh acid, it may be used in the same battery and for the same length of time. The best plan, however, is to sell it to certain branches of trade where old as well as new acids may be employed.

Finally, the acidulated water is thrown out and the jars completely emptied; the zincs are dried, after having been left a certain length of time to drip; the binding screws are dried by being put in a box of sawdust, and finally all the vessels used are emptied and cleaned.

Unless one has been in the habit of performing this long operation, he is very likely to burn his fingers with the acids; therefore the use of rubber gloves during the work is recommended.

It is a good plan also to have within reach a little ammonia, into which the fingers may be immersed in case of any accidents, or which may be put upon any spot made on the clothes by the acid.

It is seen that the use of a large Bunsen or Grove battery necessitates a long and tedious piece of work, especially if done in a new place; the work is much less, of course, in a laboratory where one has everything at hand and where the operation has been performed more than once.

If the battery is thus taken to pieces in a room, there must be some sort of an arrangement for rapidly carrying off the acid vapors; otherwise it would be impossible to finish the work or even to enter the room.

Some employ a breathing apparatus like that of M. Gallibert, with which one may remain in a place filled with a dangerous gas without any inconvenience.

In order that the zincs may be well preserved during the long intervals between experiments, they should be carefully dried and placed on end one above another, and only coming in contact at several points. As for the porous jars and the carbons, there is, as we have said, no advantage in rinsing them; it is indeed better to allow them to imbibe the acids of the battery, which gives them a more immediate conductivity in making future experiments.

COMPOSITION OF THE LIQUIDS.

We have said that, in general, dilute sulphuric acid is put with the zinc and nitric acid with the carbon; but each experimentalist has his own liquid, and some of these compositions deserve attention.

We have stated in Part I., that a solution of sulphuric acid having a density of 1.25, and composed of 30 parts of monohydrate acid for 70 parts of water, possessed the greatest conductivity, and that it is never used because too dangerous. In general, liquids containing eight or twelve parts, by weight, of sulphuric acid for a hundred of the mixture are used, whose conductivity differs but little from the maximum. In the practice, weights being more difficult to establish than volumes, the following mixtures are used: one volume of acid for six of water, or three volumes of acid at 66 hydrometric degrees for thirty-three of water.

The electro-motive force decreases (the resistance at the same time increasing) when the proportion of water is increased after the liquid has attained the density 1.25, as is shown in Table VIII. by the experiments of Paggendorf.

Generally, nitric acid at 40 hydrometric degrees is employed in the porous jars; the electro-motive force diminishes notably with the density of the nitric acid.

ACID BATTERIES. 171

Some persons, and especially Wigner, replace the nitric acid by a mixture of two parts by weight of nitric acid (specific gravity 1.360) and five of sulphuric acid (specific gravity 1.845). The proportion of nitric acid is sometimes increased to three and a half parts, if the battery is to work more than four or five hours.

It is true that Wigner's experiments were made with the Grove battery, which is only employed in England,

FIG. 40.

but it appears certain that the same results might be obtained from carbon batteries.

There is a very evident economy in the use of Wigner's mixture, as sulphuric acid costs much less than nitric acid.

In 1853 a French physicist studied this question and discovered that there was a considerable economy in substituting for the nitric acid in the Bunsen battery a concentrated solution of sulphuric acid to which was added

172 TWO-LIQUID BATTERIES.

one or two twentieths of nitric acid. Sulphuric acid evidently acts as an absorbent of water, and renders the decomposition of nitric acid much more effectual than when the latter is in a large quantity of water.

As sulphuric acid can absorb the water of its bulk in nitric acid, which is successively added in the jar, one may almost completely use up a given quantity of nitric acid, which, if used alone, would have to be thrown away long before it had become exhausted.

GERMAN MODEL.

In the beginning Bunsen placed the carbon in the outside jar and gave it the form of a hollow cylinder, in the centre of which was placed the porous jar. The porous jar contained the amalgamated zinc with acidulated water.

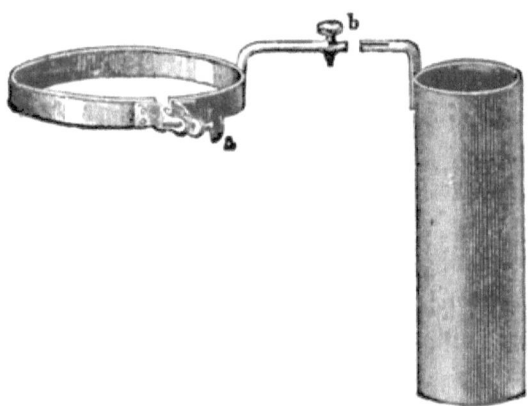

FIG. 41.

This form has been preserved in Germany and abandoned in France.

Figs. 40, 41, and 42 represent the Bunsen cell properly so called, with a hollow cylinder of rolled zinc. The

ACID BATTERIES.

glass jar is narrowed at the top in order to check the evaporation of the nitric acid.

Sometimes the hollow zinc cylinder is replaced by a rod of cast zinc. Siemens prefers a rod of cast zinc whose section has the shape of a cross.

The rod of zinc inconveniently reduces the space occupied by the dilute sulphuric acid, and consequently the quantity of the acid. The other two dispositions seem

FIG. 42.

preferable. Here arises the question, Which is the better arrangement, the German or the French?

The German model contains more nitric acid, which may increase and prolong the constancy of the battery. But we think the French model ought to be preferred, because in the German arrangement the carbon is much more expensive, the zinc is not as convenient to amalgamate on account of the strip and ring of copper soldered to it, and the expense of the acid is greater.

FAURE'S MODEL.

English books mention a form of carbon battery which deserves notice. The carbon proposed by Faure has the shape of a bottle closed with a carbon stopper. This carbon serves at the same time as porous jar and as negative electrode. It contains the nitric acid, and the vapors which free themselves force the liquid into the pores of the carbon, producing depolarization.

This form has been but seldom used. It would, however, be worthy of study, as it presents economical advantages and suppresses nitric-acid vapors, which render Bunsen's cell so troublesome and indeed dangerous for the men who have the care of a large number.

ELECTRO-MOTIVE FORCE AND RESISTANCE IN NITRIC-ACID BATTERIES.

All physicists agree that the electro-motive force of Bunsen's battery is a little less than that of Grove: the difference is very small.

As to that of Grove's battery, it varies from 1.812 to 1.512, according to the condition of the acids.

The resistance of the Bunsen is very feeble, and if the Daniell be taken as a term of comparison, it is found to vary from 4 to 10 ohms; while in the Grove battery it is less than $\frac{1}{4}$ of an ohm, at least in the beginning. At the expiration of several hours the resistance will be found to have greatly increased; but if the liquids have no longer the same composition, the battery cannot be considered as a Grove. We have often said that the resistance of cells is so variable that no precise information

ACID BATTERIES. 175

can be given. Only a general idea of the value of this resistance for each kind of cell may be given; and if one desires to replace Bunsen cells by Daniell cells, it is easy to obtain the same electro-motive force by doubling the number of cells. It is much more difficult, however, to obtain as feeble a resistance. For this electrodes having large surfaces must be employed, and they must also be placed very near each other. This is the means used by Carré, as we have shown in Chapter II.; he replaced three of Bunsen's cells by five of Daniell's. The same is obtained with Sir William Thomson's battery.

MAYNOOTH'S BATTERY.

Iron may be substituted for the carbon in Bunsen's battery without sensibly diminishing the electro-motive force. The battery may be arranged as follows:

In a cast-iron pot containing nitro-sulphuric acid—that is, a mixture of three parts by weight of nitric acid and one of sulphuric acid—is placed the porous jar, which contains the amalgamated zinc and water with one tenth sulphuric acid. The iron pot serves at the same time as negative or conducting electrode and outside jar of the battery. The advantages of this arrangement are very evident. Those who employed this battery appear to have been well satisfied with it, and we cannot imagine why it is so little used.

The part which iron takes in this arrangement has given rise to many interesting researches. It is said that the iron is rendered passive by its contact with the almost saturated solution of nitric acid.

DANIELL'S EXPERIMENTS UPON THE SIZE AND PLACE OF THE ELECTRODES.

We have explained, in speaking of single-liquid batteries, the advantage of giving a much larger surface to the negative electrode than to the positive. Those reasons do not apply to completely depolarized batteries, such as those of Daniell and Grove. It may be said that the surface of either electrode can be indifferently increased or reduced; we mean that the reasons which should lead to the arrangement of the battery are simply those pertaining to economy and practical convenience, and that the electro-motive force is the same in both instances.

Daniell has made some very conclusive experiments with regard to this subject. He constructed two Grove cells of identical dimensions. In one the soluble electrode was in the shape of a large zinc wire, and the platinum had the form of a cylinder and surrounded the cylindrical porous jar. In the other, the forms remaining the same, the platinum was placed in the centre and the zinc outside. The intensity when measured was found to be essentially the same.

Daniell repeated this experiment in diminishing the diameter of the outside cylinder without changing its height, and found the same intensity as in the first instance. Whence the very curious conclusion that the internal resistance of the cell does not alter when the diameter of the zinc cylinder alone is changed, all the parts of the cell being concentrically disposed, and the central electrode reduced to the size of a wire. The simple reason of this, which may at first appear strange, is that if the distance of the electrodes increases, the average section of the liquid increases in exactly the same proportion.

A close examination will show that this rule ceases to be true when the central electrode is of a certain size; in this case there is always an advantage in diminishing the distance between the electrodes and in increasing the dimensions of the electrode in the porous jar.

CHLORIC-ACID BATTERY.

In a series of very varied experiments a chloric-acid battery was tried. This acid, when dissolved, furnishes relatively active results, which increase in energy as the solution approaches saturation.

Although this battery can never be employed in the practice, we mention it in order to show that cells can be made after the model of those of Grove or of Bunsen.

CHROMIC-ACID BATTERY.

The battery in which the depolarizing agent is a mixture of bichromate of potass and sulphuric acid is sometimes designated by the above name. If free chromicium acid has ever been employed, it has only been in scientific experiments; it is impossible to make use of it in practical applications.

VARIOUS ACID BATTERIES.

It has been proved that hydrochloric acid has no depolarizing property, since hydrogen has no effect upon this acid.

Upon the other hand, it has been established that a mixture of nitric and hydrochloric acids has a very marked action, which is easily understood, as this mixture is a very energetic oxidant and very readily absorbs the hydrogen.

CHAPTER VI.
OXIDES IN BATTERIES.

WE have seen that oxygenated acids cause the depolarization of the conducting electrode with which it is placed by being decomposed and oxidizing the hydrogen to form water.

Analogous results may be obtained by using oxides, such as the peroxide of lead and the bioxide of manganese.

Every oxide easily decomposed—the bioxide of hydrogen, the bioxide of silver, the oxide of mercury—would give much energy to the batteries in which they are used, but the instability of the bioxide of hydrogen, as well as the cost of the others, does not permit the use of these substances in the practice.

PEROXIDE-OF-LEAD BATTERY.

About thirty years ago De La Rive constructed a battery in which depolarization was effected by means of peroxide of lead. He put the peroxide in a porous jar containing a plate of platinum, and thus obtained a battery whose electro-motive force was superior to that of the Bunsen. We believe that a plate of lead or of carbon would have done just as well, and would certainly have cost less.

Unfortunately, we do not know whether the depolarization was complete or not. We believe that by employing ordinary minium and by mixing it with pieces of crushed carbon, as we have several times recommended in the foregoing, a very economical and satisfactory battery might be obtained.

It must be noted, however, that if at the same time sulphuric acid is used with the zinc, there will be a formation of sulphate of lead, which, on account of its insolubility, might check further action. It will be remembered that the advantages of the Daniell battery consist in the great solubility of the salt formed (sulphate of zinc). Whatever may be done, and from whatever stand-point batteries may be regarded, one is always led to the choice of Daniell's sulphate-of-copper battery as a model.

PEROXIDE-OF-MANGANESE BATTERY.

At the same period De La Rive made another battery, analogous to the preceding one, by substituting peroxide of manganese for peroxide of lead. He found, however, that this battery was inferior to the preceding one.

It is certain, in effect, that the manganese battery ought to have a much smaller electro-motive force than the peroxide-of-lead battery, and that the depolarization ought to be very imperfect. Whatever may have been the reasons, this battery was completely forgotten when Leclanché commenced his researches, which resulted in the production of one of the most extensively used and best batteries, for certain instances, ever invented.

LECLANCHÉ'S BATTERY.

A cell of this battery is shown in Fig. 43. The outside glass jar is square, which allows the placing of a

OXIDES IN BATTERIES. 181

large number in a comparatively small box, thus rendering the battery less cumbersome. The glass jar is narrowed at the top, just leaving room for the cylindrical porous jar to be put in or taken out, which almost closes the glass jar, thus diminishing any possible evaporation

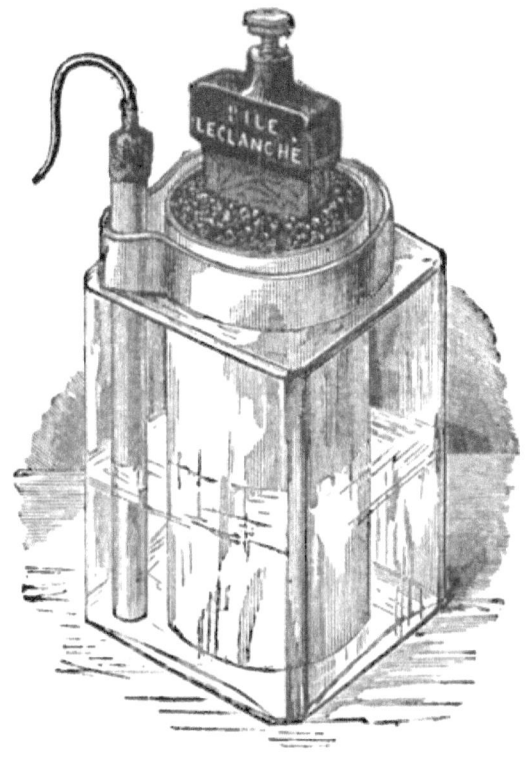

Fig. 43.

of the liquid. The narrow part of the jar is furnished with an orifice, through which the zinc is passed, and which is also convenient in pouring out the liquid contained in the jar.

The soluble electrode is formed of a simple cylindrical

piece of zinc, about half an inch in diameter. A little hole is made in the centre of the top of the zinc, in which a galvanized iron wire is soldered. This connection is at the same time flexible and solid; it may be wound in the shape of a helix, which gives it an elasticity frequently very convenient.

The porous jar has, as we have said, about the same diameter as the mouth of the glass jar, and contains almost equal parts of peroxide of manganese and crushed carbon. In the centre of this mass is a bar of carbon, capped with lead, to which the positive binding screw is attached.

The outside jar is about half filled with water and ammonia-hydrochlorate. After a short time the liquid penetrates the porous jar and enters into the mass it contains.

The mixture of peroxide and carbon is covered with wax, to prevent its spilling during transportation. There is a hole in this covering to allow the air to escape when the water penetrates the porous jar.

ADVANTAGES OF LECLANCHÉ'S BATTERY.

This battery presents many advantages, which we will enumerate:

1. The zinc is not attacked by the sal ammoniac. There is no chemical action in the battery while the circuit is open, or, in other words, there is no waste of material as long as there is no outside current produced. We have already spoken in detail upon this point in the description of sal-ammoniac batteries (single liquid); and we would only say that, from a practical point of view, this peculiarity constitutes an incontestable superiority of the Leclanché battery over that of Daniell.

2. On account of the depolarizing action of the peroxide of manganese, the electro-motive force at first starting of this cell is, as given by Leclanché himself, 1.38 (Daniell's cell taken as unit). It has indeed always been possible to replace, in the telegraph and analogous applications, a certain number of Daniell's cells by a smaller number of peroxide-of-manganese cells. Leclanché states that twenty-four of his cells can replace forty of Daniell's.

3. This battery has but a comparatively feeble resistance, resulting from the conductivity of the peroxide of manganese and of the carbon, and also from the considerable mass of the conducting electrode. In the model, in which the porous jars are $5\frac{1}{2}$ in. high, the resistance is between $5\frac{1}{2}$ and 6 units.

With equal dimensions the Leclanché has less resistance than the Daniell, which constitutes still another superiority.

It is evident that if the zinc, instead of being a rod half an inch in diameter, were rolled in the form of a cylinder surrounding the porous jar, as in the usual disposition of batteries, the resistance of the cell would be still less. As there is no consumption of zinc while the circuit is open, there is no inconvenience in increasing its surface. Consequently there is a means of reducing the resistance of a Leclanché cell, if in any particular instance it might be of advantage.

We will give, further on, the reasons which determined Leclanché in the choice of the dimensions given to the soluble electrode.

4. The battery contains no poisonous substances, neither does it throw off any acid vapors nor any appreciable odor.

5. The first cost of the materials is comparatively small.

6. The battery resists intense cold without freezing, and consequently without ceasing to work, which is proved by the following experiment:

A freezing mixture, at $-25°$ C., was placed around a Leclanché cell, in which the thermometer finally descended to $-16°$ C., without causing any appreciable slackening in the movement of an electric bell upon which the cell worked. The cell was shaken or left perfectly still, and in neither instance was any weakening (with this summary means of comparison) or tendency to freeze observed.

This battery presents, in this respect, still another notable advantage over the Daniell, which freezes in France during severe winters. The direct experiment shows that a saturated solution of sulphate of copper freezes at $-5°$ C., and a concentrated solution of sulphate of zinc at $-7°$ C.

A recent publication of Leclanché states that the resistance of his cell varies from 2.33 units at $+10°$ C. to 4.22 units at $-18°$ C., whereas that of the Daniell increases from 8.35 at $+10°$ C. to 12.58 at $0°$ C., and to 14.00 at $-4°$ C. If the temperature continues to lower, at $-6°$ C. the liquids become pasty, and towards $-20°$ C. the resistance reaches 200 units.

All this goes to show that the Leclanché battery never freezes, and that it should be used in preference to all others in all northern countries.

These advantages are of great practical importance, and explain the success of this battery, which is to-day the most extensively used for telegraphs, electric bells, and other analogous applications of electricity.

Any number of these cells may be prepared in advance and stored away without putting the liquid in them, ready for use at any moment.

After having been charged, they may be left a long time without much evaporation of the liquid and without any consumption of the materials. Their form facilitates transportation, and is capable of being modified, as will be seen, in order to obtain a perfectly closed battery. No care is needed for months at a time; it depends, of course, upon the activity of the work to be done.

The cell furnishes a more intense current than a Daniell of the same size, and almost as intense as one of Marié Davy's cells.

One must be careful, however, not to use this battery for purposes to which it is not fitted: for instance, when a continuous current or a great quantity of electricity is desired.

CONSTRUCTION AND USE.

1. We have had occasion to speak of the advantage of rolled zinc over cast zinc; we add that drawn zinc is better than either, as it is more dense and the pores are less open. Little scales are sometimes seen to detach themselves from the rolled zinc, which denotes some irregular action of the liquids upon the metal. There is nothing of the kind apparent with the drawn zinc, undoubtedly because it is more homogeneous.

We have fully explained why, in single-liquid batteries, there is an advantage in giving a greater surface to the conducting electrode than to the soluble electrode. The same reasons hold good for imperfectly depolarized batteries, such as that of Leclanché; it is seen how the inventor reduced the zinc to a small rod. We will again have occasion to speak of the dimensions of the electrode, which is quite a delicate and important question.

2. Leclanché, in recommending amalgamation of the zinc, says:

"In this battery, where there is no acid, the zinc should be used according to the theory, without amalgamation. But while the battery is at work the attack upon the zinc roughens its surface, thus facilitating the adherence of saline crystallizations when the temperature varies; whereas amalgamated zinc always presents a surface free from crystals, which fall to the bottom of the jar and do not diminish the conducting surface of zinc."

3. It is very important to employ sal ammoniac as pure as possible. That purified by sublimation is the best, although a little costly. Very good, however, can be found which has not undergone this process of purification. Care should be taken that the sal ammoniac be not dissolved in vessels of lead, for it would then contain several parts of chloride or of sulphate of lead. This condition would cause the loss of the principal advantage of the battery; for a local cell (zinc-lead) would soon form which would produce a constant and rapid waste of the zinc and sal ammoniac.

4. It is best to use an almost saturated solution; there is, indeed, no inconvenience in putting a little more salt than is necessary in the jar; it will dissolve in proportion as it is consumed by the action of the battery.

By the action of the battery salts are formed, and notably oxychloride of zinc, which is more easily dissolved in a saturated solution than in a weaker one; there is therefore an advantage in using a saturated solution, and it should not be allowed to weaken. In effect, it can be understood that if oxychloride-of-zinc crystals attach themselves to the zinc, its active surface is reduced and the resistance of the battery increased; the intensity of the battery might thus be greatly diminished. If, however, too large a quantity of sal ammoniac be added there will be the same result. This salt crystallizes upon the surface of the zinc, and the resistance of the battery becomes considerable. There can be no doubt as to the truth of this way of explaining things; for if the zincs be cleaned, the resistance will be diminished and the intensity brought back to its normal value. This observation is of great practical importance, because employés of little instruction are very apt to attribute to the battery all the faults that arise in telegraph offices, without being able to distinguish the cause; they believe the battery to be weakened, and think to restore its energy by adding more sal ammoniac; they create, in reality, the fault which they intend to correct.

5. The quality of the bioxide of manganese is also very important: that which gives the best results is the needle manganese; it is crystallized, silky, and presents a graphite appearance; if, in addition to these properties, it is hard, it possesses very great conductivity. To use it, all foreign materials must first be taken off; then it is crushed, and finally sifted to get rid of the powder. An equal volume of crushed carbon is then added. The mixture thus obtained is a very good conductor of electricity.

6. It is very important not to use powdered bioxide of manganese. The results of experiments by Leclanché show that polarization would be five times greater in a cell containing fine powder than in a cell constructed after the manner indicated above, with grains of a certain size.

The resistance of the fine powder reaches 150 or 200 units; it is considerably greater than that of the liquid with which it is dampened; consequently the hydrogen, instead of distributing itself throughout the whole mass, goes straight to the carbon plate and is not absorbed. On the other hand, the resistance of the coarser powder, as recommended above, is from 12 to 15 units inferior to that of the liquid of the battery, and consequently the hydrogen is distributed and absorbed in the whole mass.

7. The porous jar should only be half filled with the liquid. The inventor says that the dryer the matter contained in the porous jar the better the conditions of conductivity and working.

It will be seen as we advance that, by following this order of ideas, he has greatly improved his battery.

8. There is an advantage in using very porous diaphragms, in order that the action may begin as soon as the liquid has been poured in.

The quality of the porous jars is also very important. It appears that those of Wedgwood, though excellent for other batteries, are not worth anything in the Leclanché. The English have had a good deal of trouble with porous jars which chip off or burst by the solidification of the double salts of zinc and ammonium. Neither in France nor in Germany has any inconvenience of this kind arisen.

REVERSED FORM OF LECLANCHÉ'S BATTERY.

In this form the zinc is placed in the central porous jar and surrounded by the mixture of peroxide of manganese and carbon. This disposition necessitates a large quantity of manganese as compared with the sal ammoniac, and it is that which led to its being tried. There is no doubt as to the slow polarization of the battery thus arranged.

The volume of liquid, however, is much smaller than in the original form, which presents a grave inconvenience; because the liquid has thus to be renewed very frequently, and the battery loses many of its advantages. In the ordinary battery there is a quantity of manganese corresponding to the use of two zinc plates and to two charges of sal ammoniac. In the reversed form there is enough for the use of six or eight zinc plates and for twenty charges of sal ammoniac: it is a marked disproportion.

For these reasons the reversed form has obtained but little success, and should be only employed in especial instances.

AGGLOMERATED-MIXTURE BATTERY.

As the liquid has less conductivity than the bioxide of manganese mixed with carbon, it is easily understood that the resistance of the element will be diminished if the carbon electrode is well surrounded by the mixture in question, rather than by the liquid.

Leclanché observed that the conductivity increased as the matter contained in the porous jar became more compact; that is, as the empty spaces filled by the liquid became less.

In following up this idea he was led to increase the compactness to its maximum, by compressing the mixture with a hydraulic press. Porous jars became inconvenient; he suppressed them and added to the mixture a cement which held the mass together, thus constituting an agglomerate in which the carbon was tightly held which served as the conducting electrode.

The mixture is composed of 40 parts of bioxide of manganese, 55 of carbon, and 5 of gum lac, which serves to agglomerate the whole together (Fig. 44).

Finally, the inventor added in the interior of the agglomerate 3 or 4 per cent. of bisulphate of potash, which facilitates the dissolution of the oxychloride which, in the long-run, enters into the pores of the mixture.

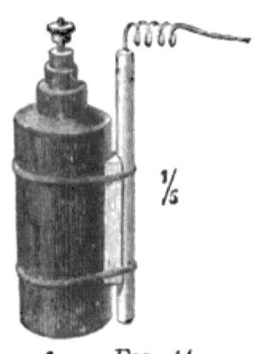

Fig. 44.

The porous jar being suppressed, some particular disposition must be adopted to prevent the zinc from touching the agglomerate, for if there were contact between them there would be the formation of a local cell and lost work. Between the zinc and agglomerate a small strip of wood might be placed and the whole be held together by two rubber bands. The zinc may also be put in a small glazed or unglazed earthenware jar whose walls are pierced with holes, which prevents the contact between the electrodes and at the same time permits a free circulation of the liquid in the outer jar. Two projecting rubber bands may be placed around the zinc, which prevent the contact between the zinc and the agglomerate.

The agglomerate, once exhausted by long work of the battery, is not worth anything; the zinc which caps the carbon may be easily taken off, as also the binding screw to which the negative connection of the adjoining cell is attached. The mass of carbon and of sesquioxide of manganese may be thrown away and the zinc and brass sold for old metals.

LECLANCHÉ'S AGGLOMERATE BATTERY.

The preceding battery has not realized the expectations of the inventor. It happened that the internal resistance of the element increased considerably, and that consequently the battery gave very unsatisfactory results, especially in circuits of little resistance.

With a view to remedying this inconvenience, Leclanché caused the agglomerate, of which we have spoken, to be made in the shape of small bricks and compressed in a hydraulic press. These little bricks are placed one on each side of the carbon electrode which rises above them; they are held in this position by rubber bands, which at the same time hold the rod of zinc and the intervening piece of wood.

"In the old battery," says Leclanché, "the internal resistance depends upon the conductivity of the agglomerated mass and upon the adherence of the carbon in this mass." The new disposition does away with the inequalities that this contact produced in the resistance, and which were the result of the production of ammonia in the interior of the agglomerate at the contact with the carbon pole. Consequently in the new battery the resistance only depends "upon the conductivity of the liquid excitant. This conductivity rather tends to increase than to diminish; in effect, by the working of the battery, chloride of zinc, which is a very good conductor, is formed; it is only the depolarizing power of the agglomerate pressed against the carbon which varies."

Besides, by increasing the number of these little bricks

placed against the carbon, the internal resistance of the element may be diminished at pleasure. Leclanché sometimes puts one, sometimes two as shown in Fig. 45, and sometimes three.

An incontestable advantage of the new battery is that

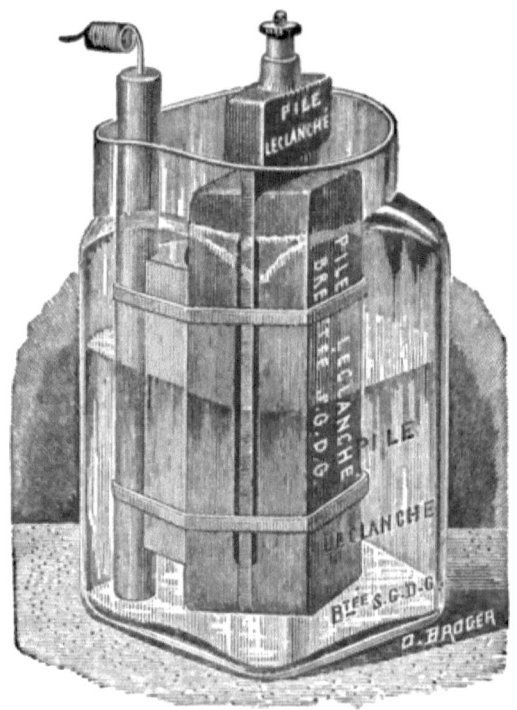

FIG. 45.

the same elements may be employed indefinitely, and it is only necessary to renew the zinc and the agglomerated bricks when they are worn.

CLARKE AND MUIRHEAD'S MODIFICATION OF LECLANCHÉ'S BATTERY.

The only difference between this battery and that of Leclanché is that in the former the carbon electrode and indeed the pieces of carbon mixed with the bioxide of manganese are platinized. This is certainly a very good idea, as we mentioned when speaking of Smee's and of Walker's batteries. The polarization is thus undoubtedly diminished.

According to information given by the inventors, this new cell, after working one minute in a circuit of 100 units of resistance, only loses 1 per cent of its electromotive force by polarization, whereas Leclanché's cell loses $2\frac{1}{2}$ per cent.

After five minutes the platinized cell only loses 2 per cent, while the Leclanché loses 5 per cent.

After ten minutes the first one again loses 2 per cent, and the second 10 per cent; if the experiment be continued, the electro-motive force of the Leclanché cell is seen to diminish steadily, while that of the platinized cell remains constant.

This platinized cell has what we have termed the reversed form; that is, the zinc is in the centre, and in a jar which is not porous, but whose walls are pierced with holes by which the communication between the two elements is established.

The zinc has the form of quite a large cylinder, the object of which is to diminish the resistance. The depolarizing mixture is placed around the sides in the outside jar, which disposition also diminishes the resistance.

The outside jar, as well as that in the centre, is closed with cement, so that all evaporation is prevented.

All these dispositions had been already tried, and the only new idea is the platinizing of the carbon.

The inventors say that they sometimes platinize the fragments of peroxide of manganese. It would be interesting to examine closely the advantages or inconveniences of this process; it might be inquired whether or not the bioxide of manganese would lose its efficiency when covered with platinum.

ELECTRO-MOTIVE FORCE. POLARIZATION.

Leclanché represented his original cell, with the porous jar, as having an electro-motive force equal to 1.38 (Daniell$=1$). These figures are certainly inferior to the real value, at least before any polarization.

We will admit the figures 1.48 as given by Clark and Sabine, which is a little less than that of Marié Davy's cell. If the circuit in which Leclanché's battery works has a considerable resistance, polarization is very slow.

CHEMICAL ACTION.

Leclanché represents the chemical action which takes place in his cell by the following equation:

$$NH_3HCl + 2MnO_2 + Zn = ZnCl + NH_3 + HO + Mn_2O_3$$

The zinc combines with the chlorine of the ammonia hydrochlorate and forms chloride of zinc; ammonia is set free; the hydrogen given off under these actions, and which would polarize the carbon without the presence of the bioxide of manganese, becomes oxidized, forming water, and the peroxide is reduced to sesquioxide of manganese.

If it be agreed to give the name of chloride of ammonium to that salt which we have called ammonia hydrochlorate, there would in reality be nothing changed, but its formula would be NH_4Cl. It could be said that the zinc is substituted for the ammonium, that the ammonium is decomposed into ammonia and into hydrogen, etc. etc.; but this manner of expression should be preferred here, for Leclanché has established that the ammonium acts more favorably in the presence of the bioxide of manganese than the hydrogen alone would; it is indeed for this reason that sal ammoniac ought to be preferred to other alkaline chlorides, such as chloride of potassium, chloride of sodium.

We ought to say, however, that this theoretical reaction in Leclanché's battery is not the only one that takes place. First, it is plain that when the battery polarizes, it is undoubtedly because the hydrogen is deposited upon the carbon and does not oxidize at the expense of the bioxide of manganese. Then is made manifest the formation of double salts, oxychloride of zinc, and double chloride of zinc and ammonium. These salts are but slightly soluble and obstruct the action. A saturated solution of sal ammoniac is needed to dissolve them, which explains one of the practical recommendations made above.

We are thus again led to remark upon the complicated nature of the chemical actions in batteries, and to say with Mr. Gladstone "that, since the elucidation of the telegraph, batteries have been studied more from a mechanical and electric point of view than from a chemical standpoint, . . . and that there is much left to be done in regard to this matter."

It is seen from the theoretical equation of the battery that ammonia is set free, and that is what the experiment shows; but in the ordinary practice of the telegraph the work is so intermittent, and the corresponding chemical actions so slow, that no odor of ammonia is noticed at all.

It should here be repeated that Leclanché did not choose sal ammoniac by chance. He tried sea-salt (chloride of sodium) and chloride of potassium, and he has shown that they are all very inferior to the sal ammoniac.

Any one can make the experiment with sea-salt and show, as we have done, that the electro-motive force of the battery thus modified is very much less than that of the Leclanché battery properly so-called, and that it polarizes rapidly. Consequently, if in any emergency the battery has to be charged with sea-salt instead of sal ammoniac, and a sufficient current is obtained, it must not be believed that a new invention has been made, nor even a good one.

WEAKENING OF THE LECLANCHÉ BATTERY.

The bioxide of manganese does not produce complete depolarization; and if the resistance of the outside circuit is very small, the electro-motive force diminishes rapidly. It is only necessary to close one of Leclanché's cells upon itself for a few seconds to polarize it in an appreciable manner. But if the current be interrupted after a short time, the battery will soon recover its initial force. This is the phenomenon of polarization in all its simplicity. If the battery only works intermittingly, as in the telegraph in general, there is very little polarization, in which case the battery is perfectly irreproachable, and deserves to be preferred to those of Daniell and of Marié Davy.

If, on the other hand, the Leclanché battery be made to work uninterruptedly, its electro-motive force is seen to decrease in a manner very interesting to watch.

Gaugain has studied this weakening under conditions well determined by measurements of the intensity of the current.

The following are the results:

	Electro-motive Force.
At the start, May 28th	288
June 1st	213
Aug. 6th	199
Sept. 2d	180
" 3d	152

This weakening would have been more rapid had the resistance of the circuit been less, and less rapid with a greater resistance.

PRACTICAL DURABILITY OF THE LECLANCHÉ BATTERY.

We have said several times that this battery possesses the great advantage of dispensing with all care during long periods, on condition, however, that it only be employed for those purposes for which it is suited.

We ought again to speak of this very important subject.

The principal French railway companies have furnished us with the following official and incontestable, though very astonishing, information:

A battery at V furnishes a continuous current, causing an electric bell to ring about twenty-three hours every day; it has needed no care during eleven months.

When the batteries at E, at Y, and at M were examined in 1876, zincs were still found which had been placed there in 1867.

Finally, a battery at O worked from July 26th, 1867, to August 12th, 1876; being recharged at that date, the zincs were renewed before there was any absolute necessity. During those nine years of service the sal ammoniac was only renewed once, and this is the only expense occasioned by that which may be considered as the average work of a railway station. This is the most striking example that has been given to us, and it leads one to believe that it would be difficult to imagine a better battery for branch offices than that of Leclanché.

It is understood, of course, that these extraordinary periods of duration are obtained by intelligent attention, or rather by the absence of any carelessness or misunderstanding. The battery at O was never touched except by the superintendent of the telegraph, who followed the experiment with great interest.

The best telegraph service is to be found in those offices where the least quantity of sal ammoniac is consumed. An intelligent inspector would always be able to distinguish the causes of the numerous hesitations in the telegraph, and act accordingly; whereas another would attribute every failing to the weakening of the battery, and would inconsiderately hasten to remedy it by adding too great a quantity of salt.

We ought here to point out the mistaken idea of the necessity of moving the battery to increase its force; some think it well to shake it in order to awaken it, as they would do a person asleep. This is a great mistake in the management of the Leclanché battery, which, it is recommended, should be kept perfectly quiet.

We have already had occasion to say that the best place

for a telegraph battery, or for one doing analogous work, is a cellar in which the temperature varies but slightly; we can only repeat this piece of advice at present. The heat in offices is rather injurious, as it occasions an active evaporation.

CHAPTER VII.
CHLORIDE BATTERIES.

THE depolarization of the conducting electrode is generally effected by oxygen, but it can also be done by chlorine, as will be seen in the batteries which we will now describe.

CHLORIDE-OF-PLATINUM BATTERY.

We only mention this battery, not admitted in practice, because Daniell points it out as a model of a perfect battery. After having described his sulphate-of-copper battery, he adds:

"The surface of the conducting electrode is thus perpetually renewed by the deposit of pure copper, and the contrary action of the zinc and of every other metal precipitated is successfully prevented. The affinity of the copper for the acid, though less, exists, however, and this opposition could not be prevented except by the use of platinum electrodes, whose surface would be continually renewed by the decomposition of chloride of platinum; this apparatus would be perfect, but very costly. . . ."

It is probable that Daniell intended the cell to be composed as follows: zinc, dilute sulphuric acid; chloride of platinum, platinum.

Under these conditions, the depolarization would be effected as in the Daniell cell, only that the hydrogen would be burnt by the chlorine instead of by the oxygen.

The principal action of the battery would still be that of the sulphuric acid upon the zinc, and the decomposition of the chloride of platinum would be less of an obstacle to the principal action than that of the sulphate of copper in the Daniell. The electro-motive force of the chloride-of-platinum battery ought to be much greater than that of the sulphate-of-copper battery.

CHLORIDE-OF-SILVER BATTERY.

Marié Davy appears to have been among the first to employ chloride of silver. In 1860 he wrote the following:

"I have constructed a battery formed of zinc, pure water, and chloride of silver melted in a silver crucible, and it has worked with perfect regularity. Its internal resistance, at first very great, has gradually diminished in proportion as the chloride of zinc formed has been dissolved in the water. By previously dissolving this salt, the battery immediately furnishes a strong current. The chloride of silver is completely reduced throughout all its parts, always preserving its shape. The insolubility of the reducible salt becomes an advantage, as it dispenses with the use of porous jars. . . ."

During the same year we studied this battery, using a porous jar and undissolved chloride of silver; our electrodes were of copper and amalgamated zinc. We found the electro-motive force to be apparently equal to that of the Daniell.

These experiments possess in themselves but little interest. It is only since Warren De La Rue has given his attention to this battery that it has attained any importance, and that its use has become general.

In the beginning of his researches, this physicist used chloride of silver in the shape of a powder or paste, and his liquid was a thin solution of sea-salt. A little inconvenience in this battery has been pointed out to us. It appears that it evolves gas, and that, consequently, if the glass jar be hermetically closed the pressure of the gas causes it to burst.

It is clear that this gas is only an inconvenience when the jars are tightly closed, and that its gravity should not be exaggerated.

At all events, we believe that the form which De La Rue has given to the battery does not present this little disadvantage, and that it contains several very interesting dispositions. Figure 46 represents a battery of ten cells, each of which is composed in the following manner:

The outside cylindrical jar is about 5 in. high and 1½ in. in diameter. The soluble electrode is formed of a rod of unamalgamated zinc, but of very good quality. In the upper part of this zinc rod is bored a hole, in which a strip of silver (positive connection of the adjoining cell) is held by means of a small brass wedge, which assures a perfect contact between the zinc and the silver.

The other electrode is formed of a strip of silver around which is melted a cylinder of chloride of silver, AgCl, represented separately in the figure.

To avoid an accidental contact of the two electrodes, the rod of chloride of silver is placed in a small cylinder of parchment paper, A; there are two holes near the top of this paper cylinder, through which the strip of silver passes, as shown in B.

The liquid is a diluted solution of sal ammoniac; the best proportion is that of 23 grammes of chloride of ammonium to 1 litre of distilled water.

CHLORIDE BATTERIES.

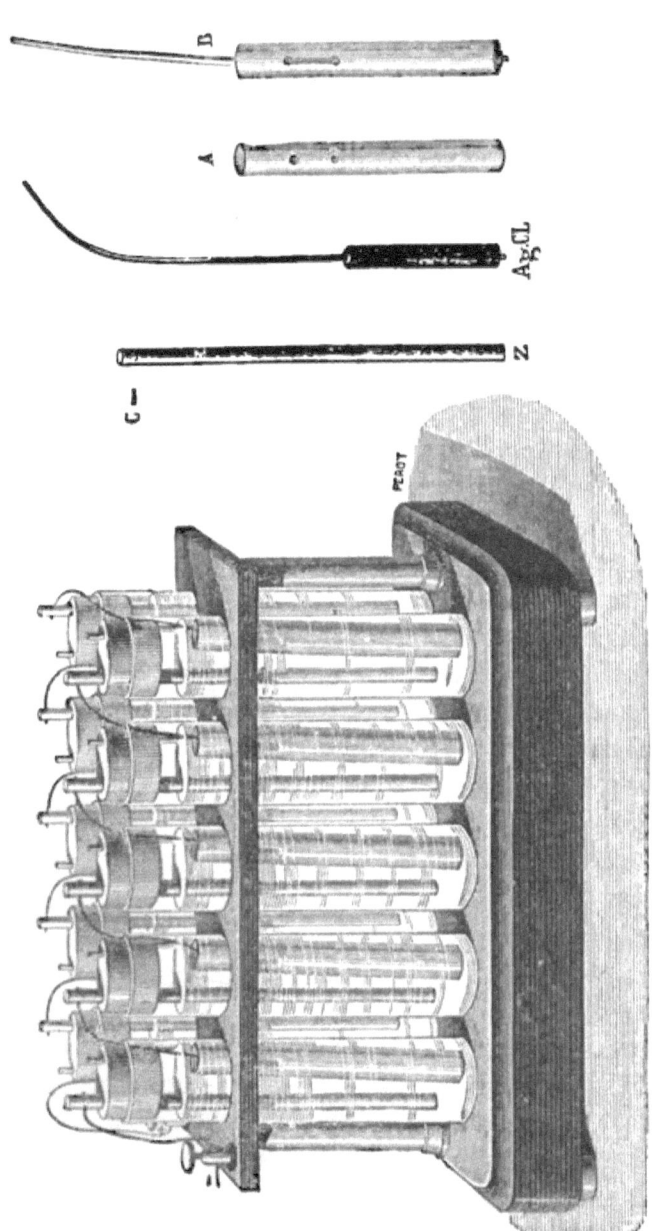

FIG. 46.

The outside jar is closed with a paraffin stopper through which the zinc passes; the strip of silver passes between the stopper and the glass jar. The figure also shows that there is a hole made in the paraffin stopper, through which the liquid is poured in by means of a small funnel, and then the hole is closed with a small paraffin cork.

The choice of this material presents important advantages. First, paraffin is one of the best insulating substances, which is of great importance when several thousand cells are joined in intensity, as De La Rue does. Then it is absolutely antihygrometric, so that if a little water be dropped upon its surface it collects in many little globules and does not spread over the surface, which might destroy the insulation. The air does not deposit itself upon it as upon glass, for instance, whose surface thus covered with steam becomes a conductor. Finally, paraffin melts when slightly heated. The jar may be hermetically closed by means of a small iron wire, flattened at the end and slightly heated, which is passed around the zinc where it comes through the cork, and also around the paraffin stopper which closes the orifice for the introduction of the liquid.

The action in this battery is very simple: the zinc is dissolved and takes the place of the silver in the chloride; the silver is deposited in a porous mass first upon the surface and then in the mass of the chloride.

The capital advantage of this battery, as in all where zinc with sal ammoniac is used, consists in the absence of any local or internal action as long as the electric circuit is open; in other words, this battery does not work upon itself. This circumstance is very important, for these cells are only used in experiments of short duration and long intervals; it is necessary to find the battery, eight days later, in exactly the same state in which it was left eight days previous.

In the first moments of the action the current is very feeble; this is caused by the great resistance, for the conductivity of the chloride of silver is very little and the exposed surface of the silver electrode is very small. But at the end of a certain time the surface of the chloride is covered with silver, its conductivity is increased, and the intensity attains its normal value, which varies but little afterwards, as is shown by the following table of experiments made by De La Rue with a voltameter:

	Gas per Minute.
At the start, June 29th, 1875	1 cent. cube.
July 4th, "	1.4 " "
Oct. 27th, "	1.4 " "
March 15th, 1876	1.45 " "
April 8th, "	1.41 " "

The electro-motive force of these cells differs very little from that of the Daniell. De La Rue found it to be 0.97 with sea salt, and 1.03 with sal ammoniac.

The resistance is about 4.3 ohms, as nearly as could be ascertained.

For the above-mentioned experiments De La Rue collects 200 cells upon a single board; they pass through a slab of ebonite, which is a better insulator than wood.

Six of these batteries of 200 cells each are placed one above the other in a carefully constructed closet which protects them from dust or from accidents.

This is not the place to enlarge upon De La Rue's experiment. We would only say that he has again proved that electricity generated by batteries differs in no way from that generated by electric machines, and that if a sufficient number of cells be joined to form a battery, a continuous spark is observed when the extremities of the two connections are brought in close proximity to each other. De La Rue has collected as many as 8000 cells in one battery, many more than any one had previously collected, and he proposes soon to make a battery of 12,000 cells.

There can be no polarization in the experiments made by this physicist, as they last but a short time and are made at such long intervals, as is the case in nearly all purely scientific experiments. But Du Moncel has made some direct experiments which prove that after the circuit has been closed twenty hours the battery shows no polarization; and if, in fact, the action takes place as we have said—that is, the simple substitution of zinc for silver in the chloride—there is no hydrogen set free, and consequently no polarization. As the deposit of silver upon the conducting electrode does not change the nature of the latter, the conditions are the same as in the Daniell battery, which is the model of completely depolarized batteries.

GAIFFE'S BATTERY.

The chloride-of-silver battery is very extensively used by Gaiffe for induction-coils or to furnish continuous currents used for medicinal purposes.

His cells are very small and hermetically closed in ebonite boxes having screw tops.

In those batteries which are destined to be transported frequently from one place to another there is no free liquid; the two electrodes are separated by six or eight sheets of blotting-paper saturated with a solution containing 5 per cent of chloride of zinc.

CHLORIDE BATTERIES. 207

Gaiffe has employed powdered chloride of silver, but he now seems to prefer the melted chloride.

The residue of the battery is silver, and if it be kept and given back to the manufacturer its use is very economical.

Figure 47 represents another cell which stands on end, and in which liquid is placed as in ordinary batteries.

The electrodes are, as is seen, attached to metallic pieces

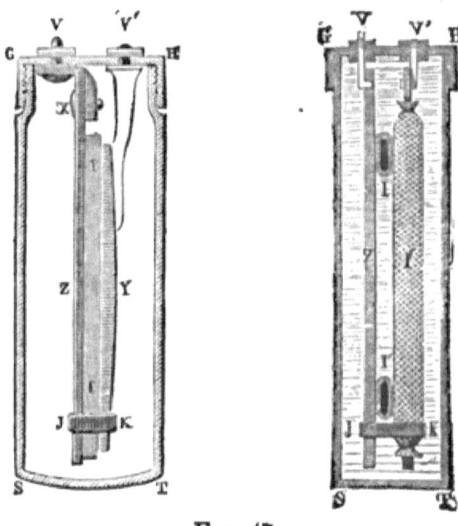

FIG. 47.

which pass through the top, and to which are fastened the connections of the adjoining cells.

There is no waste in these batteries when the circuit is open, which is of great advantage. It is important to note that, in order not to lose this advantage, the top must be kept perfectly dry, for the least humidity that might join the electrodes would establish a circuit and produce a constant working in the cell.

CHLORIDE-OF-LEAD BATTERY.

Marié Davy tried the use of chloride of lead, but the electro-motive force thus obtained was less than the unit (Daniell's cell). There is therefore no advantage in its use, for chloride of lead is comparatively dear.

PERCHLORIDE-OF-IRON BATTERY.

In 1866 Duchemin proposed the use of perchloride of iron as depolarizing agent. This substance is pointed out as containing a quantity of chlorine, as are the peroxides of lead and of manganese, which contain large quantities of oxygen.

The battery in question has an electrode of zinc immersed in a solution of sea-salt, and an electrode of carbon in a solution of perchloride of iron.

The hydrogen given off by the action of the zinc upon the water goes straight to the carbon and decomposes the iron salt, which is transformed into protochloride.

The hydrochloric acid, formed by the combining of the polarizing hydrogen and the chlorine set free from the perchloride, contributes to the dissolving of the zinc and to the intensity of the battery.

This cell is not constant; first because depolarization is not complete, and again because there are deposits, possessing little conductivity, made upon the zinc.

Du Moncel has shown that the electro-motive force of this battery is, at first starting, very superior to that of Marié Davy's battery, and inferior to that of Bunsen's battery.

CHLORIDE BATTERIES. 209

The figures are:

    Bunsen or Grove cell...........................11,123
    Perchloride of iron............................ 9,640
    Marié Davy's sulphate-of-mercury cell.......... 8,192

The contrivance proposed by Duchemin would therefore appear worthy of study. It is probable that, by means of some chemical or physical artifices, the constancy of this battery might be increased.

The mixture of pieces of carbon with the perchloride of iron has already been tried, and with good results.

CHAPTER VIII.

DEPOLARIZING-MIXTURE BATTERIES.

We have seen in the previous chapters that the constant preoccupation of physicists should be to depolarize the conducting electrode by surrounding it with those bodies from which oxygen or chlorine is easily freed. These gases combine with the gases evolved by the action of the battery, and prevent or diminish the polarization of the electrode. Instead of employing bodies which furnish oxygen or chlorine by their decomposition, a mixture of two substances, whose reciprocal reaction produces oxygen or chlorine, may be placed around the electrode.

A battery may therefore be made by using any of the means indicated in works upon chemistry for the preparation of oxygen or chlorine.

We will examine the most important of these batteries, commencing with those in which oxygen is the depolarizing agent.

POTASSIUM-CHLORATE AND SULPHURIC-ACID BATTERY.

In 1859 Messrs. Salleron and Renow presented to the French Academy of Sciences a battery in which depolarization was effected by means of a mixture of potassium chlorate and dilute sulphuric acid.

This battery appears not to have been used except in a few experiments made by the inventors. Its failure, however, may have been due to some secondary cause, and we believe that the idea might be carried out, and with very good results.

The electro-motive force of this battery is less than that of Grove's cell, but greater than that of the Daniell.

The inventors call attention to the fact that the potassium chlorate destroys five times the quantity of hydrogen that sulphate of copper does, and that its cost is only about three times greater, whence their justifiable conclusion that the battery ought to be economical.

Unfortunately we have not been able to ascertain whether the depolarization is complete or not; we have not had time to make the experiment ourselves.

ANALOGOUS CONTRIVANCES.

Potassium chlorate could be replaced by chlorate of sodium, and the chlorates by the nitrates.

The mixture of nitrate of sodium and sulphuric acid has been tried, and it presents several economical advantages, because nitrate of sodium is a very cheap salt found in large quantities in South America.

We do not believe that this battery has ever been used in the practice, but we do not know what disadvantages it might possess.

BICHROMATE-OF-POTASSIUM AND SULPHURIC-ACID BATTERIES.

Among the many known means for the preparation of oxygen, there is one which consists in putting bichromate of potassium and acidulated water in a retort. The reaction is indicated by the following equation:

$$KO.2CrO_3 + 4SO_3 = Cr_2O_3, 3SO_3 + KO.SO_3 + O_3$$

That is, an alum (double salt having the formula $Cr_2O_3, 3SO_3, + KO.SO_3$) and oxygen are formed.

The use of this mixture for depolarizing the conducting electrode was an idea of Poggendorff, who thus conceived a very interesting voltaic combination, which has received many applications under the great variety of forms given to it by the constructors.

The most simple form of this cell is the same as that of the Bunsen: amalgamated zinc in a glass or earthenware jar, porous jar in the centre of the zinc cylinder, carbon in the porous jar. In the outside jar is sulphuric acid diluted with twelve times its weight of water. In the porous jar is a mixture composed as follows:

      100 parts of water.
       12 " " bichromate of potassium.
       25 " " sulphuric acid.

This battery thus charged has an electro-motive force greater than that of any battery that we have as yet studied; it is equal to 2.028 volts, according to Clark and Sabine, which means that it is double that of the Daniell cell.

It is true that W. H. Preece, Electrician of the Post-office, England, found it to be equal to 1.97 at its maximum, with a liquid composed a little differently, of which we will speak as we proceed. But even admitting this latter value, it is seen that the electro-motive force of the battery in question is greater than that of either Grove's or Bunsen's battery.

It must be added that this extraordinary electro-motive force is only realized at first starting, for the battery polarizes rapidly, at least if placed in a very short circuit.

The conclusion to be drawn from this last observation is that Poggendorff's depolarizing mixture accomplishes but incompletely its object; for this reason his idea, however ingenious it may be, cannot be placed side by side with the inventions of Daniell or Grove.

It is important to note that the two substances of which the mixture is composed act upon each other independently of any action of the battery; consequently the liquid ceases, after a certain time, to possess any depolarizing virtue. In 1841 Bunsen suggested a much more complex mixture of chromate of potash, chloride of potassium, bioxide of manganese, and common salt. Wiedemann says that this mixture gives less satisfactory results than that suggested later by Poggendorff.

CHEMICAL ACTION IN THE BICHROMATE BATTERY.

Poggendorff gives the following as the composition of the mixture to be placed around the carbon:

Bichromate of potash	3 parts.
Sulphuric acid	4 "
Water	18 "

Another composition, as suggested by Wöhler and Buff, is:

Bichromate of potash	12 parts.
Sulphuric acid	25 "
Water	100 "

It is with this latter mixture that the measurements for the electro-motive force stated above were taken, and it is that which is used in the German telegraphs.

Many other proportions have been suggested, and there

prevails concerning this subject a certain confusion which we will endeavor to do away with.

It should be well understood that we are now speaking of batteries of two liquids, separated by a porous partition.*

Returning to the equation representing the nature of the chemical action produced by the reaction of the sulphuric acid upon the bichromate of potash, it is seen that four equivalents of sulphuric acid must be made to react upon one of bichromate.

Bichromate of potash is an anhydride, whose equivalent is:

$$KO \ldots\ldots\ldots 47.11$$
$$2CrO_3 \ldots\ldots\ldots 100.56$$
$$\overline{147.67}$$

Sulphuric acid, the monohydrate, has the following equivalent:

$$SO_3 \ldots\ldots\ldots 40$$
$$HO \ldots\ldots\ldots 9$$
$$\overline{49}$$

and four equivalents have a weight equal to $49 \times 4 = 196$.

Thus the theoretical proportion is that of 147 to 196, or in round numbers 150 to 200, which is that of 3 to 4 as indicated by Poggendorff. We repeat once more that it is the sulphuric monohydrate that we speak of. If ordinary commercial sulphuric acid were used, the proportion of 3 to 4 would no longer be in accordance with the theory (which we have just exposed).

---

*We will speak later of bichromate batteries without porous jar, which are termed (wrongly, according to our idea) single-liquid batteries. The composition of the liquid ought to be different.

It must be admitted that the proportion as indicated by Poggendorff has only a theoretical value, because the commercial sulphuric acid which is necessarily employed in the practice is not a monohydrate, but contains more water.

It is probable that Wöhler and Buff increased the proportion of sulphuric acid because they found that an excess of the acid facilitated the action; perhaps they also observed that the conductivity of the liquid was greater with the composition they suggest.

If we were studying batteries without porous jar, the proportions ought to be very different; the reaction may be represented by the equation:

$$3Zn + KO.2CrO_3 + 7SO_3HO = 3ZnO.SO_3 + Cr_2O_3 3SO_3 + KOSO_3 + 7HO$$

Therefore, one equivalent of bichromate of potash (147.67) and seven equivalents of sulphuric monohydrate ($49 \times 7 = 343$) are theoretically necessary, say in round numbers 3 to 7.

Mr. Byrne recommends the following mixture, in which the proportion of sulphuric acid is still greater than that indicated by the preceding calculation:

    340 grammes of bichromate of potash.
    925 " " sulphuric acid.
    2500 " " water (2½ litres).

The proportion of 925 grammes of acid to 2500 litres of water is greater than that which corresponds to the maximum of conductibility of the mixture (see Table IV. at the end of the book). But it must be taken into consideration that, as soon as the battery begins to work, the acid weakens; therefore it is quite justifiable to put a larger quantity than is necessary at first starting.

APPLICATION TO THE TELEGRAPH.

The bichromate-of-potash battery is employed in Prussia in very important offices. Its considerable electro-motive force has caused it to be preferred to the Daniell (Meidinger's balloon form) in all places where there is an especial person to take care of it, and where the handling is not of such great importance.

On account of its great electro-motive force and its feeble resistance, this battery is well suited to those offices where work is done on several lines at once with a single battery.

"The necessity of frequently renewing the elements," says Dr. Dehms, "is an inconvenience, but there is no difficulty in the operation. 'The Commission' has in part avoided this inconvenience by proposing the construction of very large cells. The large quantity of material that they are able to hold necessitates a less frequent renewal.

"The electro-motive force and the resistance vary in an unfavorable manner from the beginning until the battery is completely exhausted. The electro-motive force diminishes but slightly, and the resistance increases considerably. In order to obtain a greater constancy, a certain number of elements may be changed at regular intervals, and not the whole battery at one time."

The elements used by the German Government are disposed as has been described when speaking of the German model of Bunsen's battery.

The carbon, having a height of $6\frac{1}{3}$ inches and an exterior diameter of $3\frac{1}{4}$ inches, is placed in the outer jar, around the porous jar; its walls are pierced with nine holes at different heights, in order to facilitate the circulation of the liquid. A copper ring is closely laid around the upper part, which has previously been immersed in a bath of paraffine, as we have explained in speaking of other batteries, to prevent the attack on the metal by the acids; this ring is separated from the carbon by a thin strip of tin.

The zinc is placed in the porous jar and is cross-shaped, which gives a considerable surface with a comparatively small mass.

The liquids used are: with the zinc, sulphuric acid diluted with 20 times its bulk of water; and with the carbon, one part by weight of bichromate of potash, two parts of sulphuric acid and eight of water, the latter liquid being the theoretical mixture indicated by Buff.

We believe it is well in this battery to give a large surface to the carbon, and to place it in the outer jar; the battery not being completely depolarized by the mixture, there is an advantage, as we have several times repeated, in increasing the size of the conducting electrode. As the liquids are not subject to evaporation, there is no inconvenience in increasing the quantity of that liquid which produces the depolarization. The disadvantages of the cast-zinc disappear with the amalgamation, at least if the amalgam is renewed frequently enough.

Dr. Dehms calls this a Bunsen battery, with chromic acid. We believe, however, basing our opinion upon the authority of Wiedemann, that this is a misnomer.

GAUGAIN'S EXPERIMENTS.

Gaugain has given the results of several very interesting experiments with bichromate-of-potash batteries.

He has shown that chrome alum forms in the battery even when not at work, but that the electro-motive force is only slightly diminished by this alteration in the liquid; it only varies from 296 to 278 in four months (open circuit).

He found that by causing the battery to work upon an electric bell night and day, the electro-motive force was, after 17 weeks' work, still superior to that of the Daniell cell.

These figures show that the battery in question, although less constant than the Leclanché, is still a good battery, especially when a large number of cells are to be collected together in a limited space. It will furnish an equal current with fewer cells. There will be no freeing of gases or odors, and it will only require more frequent attention.

USE IN ENGLAND.

FULLER'S BATTERY.

The bichromate-of-potash battery has also been employed in England since 1877, under a form given to it by Fuller. The zinc plate is placed upright in the porous jar, and held in this position by means of a kind of ring at the bottom. It should be carefully amalgamated, as also the copper wire which rises from the bottom and serves as the connection. Finally, an ounce of liquid mercury is put in the bottom of the porous jar, as we have seen done by Tyer and other electricians.

The carbon does not surround the porous jar as in the German model; it is a simple plate 6 in. by 2 in., and provided with a metallic cap to which a clamp screw is attached.

We owe to Mr. Spagnoletti some very interesting information upon the results obtained from the use of this battery upon the Great Western Railway, England.

He estimates its electro-motive force at 2 volts and its resistance at 1 ohm, for the model whose capacity is one litre. At Paddington Station the service, directed upon 11 lines, varying from 42 to 284 miles, is done at present with 64 of Fuller's cells only.*

This battery worked actively night and day the whole of the year 1878, during which time sulphuric acid was added ten times and bichromate only five times. At the end of December the cells were dismounted and thoroughly cleaned; several zinc plates had to be replaced.

It is seen that the care of the battery is reduced very much; and indeed during the first three or four months no attention whatever is necessary.

Mr. Spagnoletti considers this battery as very convenient for branch offices; and indeed, when compared with the Daniell, the reduction to one half of the number of cells is no small advantage.

The cost of keeping the battery in order is not very great, as the carbon ought to last a long time, and the zinc from twelve to eighteen months; the sulphuric acid and the bichromate are substances which cost very little.

It appears that there are already 20,000 cells in use in England, which is proof of a great success. We know from

---

* It is interesting to note that at Paddington Station 64 of Fuller's cells replace 575 of Daniell's of the model shown in Fig. 23. This enormous progress is, however, not wholly due to the superiority of the bichromate battery; it is due in a great measure to the adoption of the "Universal System."

good authority that 3000 cells are now in use at the General Post-office, London.

The only new peculiarity we notice in Fuller's battery is the amalgamation of the zinc, which suppresses all local actions and prevents the waste of the zinc during the whole time the circuit is open. This is a capital advantage in its application to the telegraph.

This battery appears to us to be well suited to important telegraph offices where the Leclanché is out of place. It also does very well, as Mr. Spagnoletti says, for closed current services.

For ordinary telegraph offices, however, the Leclanché, or the chloride-of-lime battery, appears to us to be superior. It is true that a larger number of cells is required, but the handling of any acid is dispensed with, and the battery may remain many months without being visited, and years without renewing the zinc.

MILITARY BATTERIES.

Bichromate-of-potash batteries with especial dispositions are used to set fire to military mines. A description of them can be found in Captain Picardat's book.* We will only describe two of these apparatus.

The single-cell battery, represented by Fig. 48, is composed of a hollow zinc cylinder, in the centre of which is a rod of carbon. The outer surface of the zinc, which is not designed to effectually contribute to the useful action of the battery, is painted over with a black varnish, which shields it from the attack of the liquid. The two electrodes are fastened to a small circular piece of wood,

---

* "Les Mines dans la Guerre de Campagne." Picardat, 1874.

DEPOLARIZING-MIXTURE BATTERIES.

which is provided with terminals to which the conductors are attached. The liquid is contained in a glass jar closed with a wooden tampion surrounded with rubber.

The electrodes are only immersed in the liquid at the precise moment when the mine is to be exploded, and they only remain a few seconds. Under these circumstances the battery has its maximum effect.

If, in a very short time after use, the electrodes can be washed in pure water, there will be no sensible waste of the zinc; neither will any liquid have been consumed.

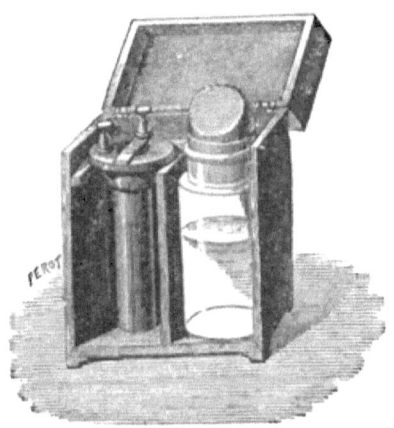

FIG. 48.

Consequently with a little care and attention the battery may do service a long time without being renewed or repaired. The whole apparatus is enclosed in a small box with two compartments, which the figure plainly shows. This simple battery is known by the name of "Arras," because it was first used at the *École régimentaire du Génie d'Arras* by Captain Barisien. In fact this battery was originally arranged as we will now describe.

During the siege of Paris, in 1871, batteries of four

cells were constructed. Each cell is composed of a half cylinder of zinc and a half cylinder of carbon, separated at the upper part by a little piece of ebonite. They are supported by being fastened in holes made in a small board; these holes may be lined with copper, which establishes a good contact between the electrodes and the connections of the adjoining cells. A wooden handle allows the four cells to be moved at once and simultaneously immersed in the jars containing the liquid. There was in the box a second compartment where the electrodes remain separated from the liquid while the battery is not at work.

We have described this apparatus to show especially how it may be improvized, perhaps not in the country, but in a besieged town or city, where there might be a lack of more perfect material.

GRENET'S BOTTLE BATTERY.

This battery (Fig. 49) has the form of a bottle with a wide mouth, the lower part being almost spherical. The top is provided with a brass frame, to which is fastened an ebonite cover. To this cover are attached two carbon plates which permanently dip into the liquid, and which are held apart at the bottom by means of small pieces of ebonite. Between the carbon plates is suspended a zinc plate which may be plunged into the fluid or withdrawn at pleasure. The contact between the carbon and the zinc is prevented by little pieces of ebonite fastened to the carbon and which serve to guide the movement of the zinc.

This battery is employed very extensively in laboratories, and presents some very great advantages: 1st, the resistance is very slight, on account of the short distance between the electrodes; 2d, the waste of the zinc is suppressed during the intervals between experiments, as it is withdrawn from the liquid; 3d, polarization is slackened by the comparatively large surface of the carbon electrode; 4th, the quantity of liquid is considerable on account of the spherical form of the lower part of the bottle; 5th, and finally, the charging and cleaning of

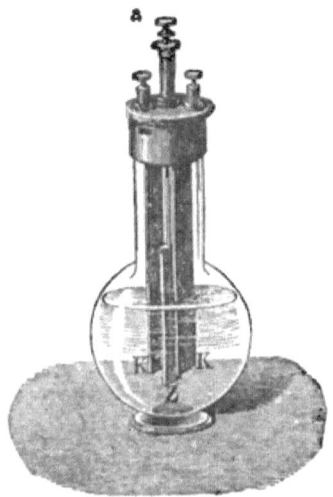

Fig. 49.

the battery is extremely easy, as there are but a single jar and a single liquid.

In spite of these advantageous dispositions, the battery gives a powerful current only for a short time, after which the intensity is seen to diminish. This element is therefore only suitable for experiments of very short duration, such as are made in laboratories, or for surgical operations which last but a few minutes.

Sometimes this element is complicated by using three

carbon plates and two zinc plates; the surface of the electrodes is thus increased, and the battery still consists of one cell. A cell of this kind is frequently employed, alone or joined with others identical, to excite Rhumkorff's induction-coils.

Grenet used a leaden tube starting from the cover and going to the bottom of the liquid. This tube served to introduce air into the bottle and to agitate the liquid. This idea has been abandoned in ordinary practice, although it possessed great merits.

Although this be an element with no porous partition, it is sometimes wrongly called a single-liquid cell; there are in reality two liquids mixed; viz., sulphuric acid designed to act upon the zinc, and a mixture of acid and bichromate intended to depolarize the carbon.

TROUVÉ'S BATTERY.

This battery, represented by Fig. 50, is a derivative of the preceding one. A certain number of zinc and carbon plates are placed in a suitable ebonite frame, and are held at short and equal distances apart, in such a manner as to be joined to form either a single element with a large surface or two elements with a smaller surface which are joined in intensity.

The frame or box in which the plates are placed is formed of an ebonite base and two vertical ebonite supports, N, united and held at the top by the handle, A. The distances between the plates are maintained by bands of rubber placed around the carbons horizontally. In case of any accidental shaking these rubber bands deaden the shock, which is very important for the carbons, as their fragility is such as to necessitate great care to prevent their being broken. Metallic and movable clamps are placed upon the zinc and carbon plates, and are attached to cross-pieces above, which join several zinc or carbon plates together.

In the figure, to the right, in front, are seen three carbon plates joined together; this is the positive pole of the

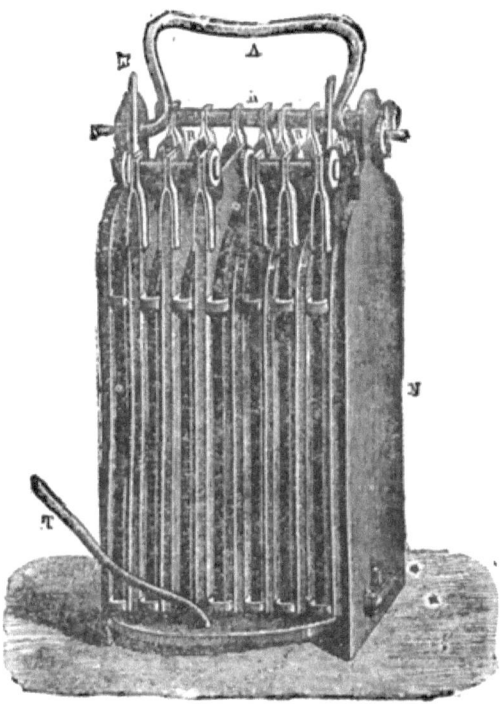

Fig. 50.

battery. The three corresponding zinc plates are behind, joined to each other and to three other carbon plates; finally, the last three zinc plates are united to the left, in front, and this forms the negative pole of the battery, which is thus composed of two cells joined in intensity.

The two cells are immersed in a single trough containing the liquid, or rather the mixture. There is, to be sure, a certain loss of current by the liquids, but the simplicity of the battery more than compensates for the fault in question.

A tube, T, permits the introduction of air, which goes to the bottom of the liquid, agitates it, and contributes to the depolarization. The battery may be shaken in the liquid by means of the handle, which will produce almost the same effect as the injection of the air.

It is seen from the preceding that Trouvé's arrangement presents the following advantages:

1. The battery can be easily and rapidly dismounted, a convenience wanting in Grenet's element.

2. The plates, when dismounted, can be conveniently washed, which prevents any slow waste caused by the acids.

3. The zinc plates can be reamalgamated, and when worn can be replaced without the help of a special constructor.

4. The clamps washed and dried may serve indefinitely.

Finally, the battery may be so arranged as to form two or more elements, or a single cell.

BYRNE'S PNEUMATIC BATTERY.

Dr. Byrne, of Brooklyn, New York, has invented, upon the preceding principles, a battery which attracted a great deal of attention in 1878, and which deserves to be dwelt upon.

The positive electrode is formed of a plate of zinc and placed between the two negative electrodes, as in Grenet's battery.

The double negative electrode is in reality a very thin plate of platinum. But Dr. Byrne, taking into account the insufficient conductibility of this thin plate, faced one side of it with a plate of copper. Moreover, to avoid the attack upon the copper, he coated it with lead. Thus the electrode consists of a copper plate which is coated with lead, and which has one side faced with a plate of platinum. A layer of varnish protects the parts of the lead which are not covered by the platinum.

The element is placed in a large square ebonite jar, provided with a cover of the same material, to which the two negative electrodes are attached. Between them is suspended the zinc plate, which may be immersed in the liquid or taken out when the battery is not at work.

Finally, Dr. Byrne affixed to his battery a means of aërating or agitating the liquid. This apparatus consists of a perforated rubber tube, which is fixed in the bottom of each cell, and through which air can be forced into the solution by means of a fan or blower; this is what Grenet did as early as 1857.

The intensity of the current furnished by this element is considerable; it is especially suited to medical uses, and particularly to cautery purposes.

With a battery of 10 cells a stout platinum wire, 30 inches long and $1\frac{1}{2}$ inches in diameter, was brought to a glowing red heat.

The advantage of the injection of air is shown by the fact that when air is being forced in the platinum wire is quickly brought to a glowing red heat, and that on ceasing to inject air the wire gradually cools down.

The electro-motive force of this element varies but slightly (from 1.73 to 1.97), as will be seen further on; but its variable resistance (from 0.78 to 0.14 ohm) is always exceedingly small.

This feeble resistance is the result of several circumstances:

1. The nature of the negative electrode, which is very well combined, and is superior to the carbon electrode.

2. The great conductibility of the liquid, which is very rich in sulphuric acid.

3. The high temperature which is produced when the circuit is closed and air injected.

AGITATION OF THE LIQUID.

We have already spoken of the advantage of agitating the liquid or the electrodes of a battery subject to polarization. The result is a freeing of hydrogen bubbles which diminishes the polarization. It happens also, sometimes, that precipitates possessing little conductibility form themselves upon the conducting electrodes, and which the agitation causes to fall to the bottom of the jar.

This latter impediment is met with in the bichromate battery; it happens in certain instances that there is formed, besides the chrome alum, a yellow precipitate which partially covers the carbon and diminishes the intensity of the battery.

From the beginning, Grenet, to whom is due the very excellent form of the element that we have made known, arranged a tube of lead descending to the bottom of the liquid; to this tube was fastened another one of rubber, through which air was injected, either by blowing with the mouth or with a bellows. This air passed through the liquid from top to bottom, agitated it and depolarized the battery: at least that is what it was believed to do at the time.

Preece has lately made some very extensive experiments upon the effect of the injection of air in Byrne's battery, which have given fresh interest to the subject.

Ladd announced that by introducing successively into the battery common air, oxygen, and hydrogen, no difference was observed. Preece confirmed the truth of this statement, and obtained the same effects by injecting a liquid into the liquid. It is therefore well established that the air does not act chemically, but purely by the agitation it produces.

Preece has also established that the electro-motive force, though not invariable, varies but very little, the limits being 1.73 and 1.97. The result is that the cause of the change in the intensity of the current is not depolarization in the ordinary sense of the word.

His experiments have proved, moreover, that the internal resistance of the battery varies considerably as the temperature of the liquid is increased under the influence of the agitation.

This is shown by the subjoined table:

Temperature.		Electro-motive Force.	Resistance.
Fahr.	Centig.	(Daniell=1.)	(Ohm=1.)
80	26.66	1.73	0.78
100	37.77	1.88	0.61
120	48.88	1.92	0.35
140	60.00	1.97	0.24
160	71.11	1.97	0.19
180	82.22	1.97	0.17
200	93.33	1.97	0.14

In order to further elucidate the question, Preece submitted the liquid to a direct experiment, outside of the battery and away from the zinc; he employed two platinum electrodes, as is almost always done when the resistance of liquids is to be ascertained, and he found the following figures:

TEMPERATURE.		RESISTANCE IN OHMS.
Fahr.	Centig.	
70	21.11	1.78
90	32.22	1.50
110	43.33	1.37
130	54.44	1.15
150	65.55	1.00
170	76.66	0.79
190	88.88	0.43
210	99.99	0.20
212 (boiling point)		0.04 (variable)

But the high temperature is not developed by the action alone of the acid upon the bichromate of potash; the zinc must necessarily be present. One is led to the belief that the circulation of the liquid promoted by the air tends continually to bring fresh acid in contact with the zinc plate and to disperse the salt formed at its surface, which would obstruct further action of the acid.

This circulation of the liquid is especially necessary to cause a renewal of the contact (between the liquid and the positive electrode) when the electrodes are very near each other, as is the case in the batteries of Grenet, of Trouvé, and of Byrne, because in this latter instance the natural motion of the liquids is slower and more difficult.

Other inventors, seeing the utility of the agitation in the liquid, have sought to improve upon the process suggested by Grenet. They have (first Chutaux and then Camacho) arranged several elements in gradation, the liquids flowing from one to the other and thereby producing a renewal of the active fluid at the contact with the electrodes.

CAMACHO'S BATTERY.

We have said above that Chutaux sought to produce the agitation of the liquids which is so advantageous in

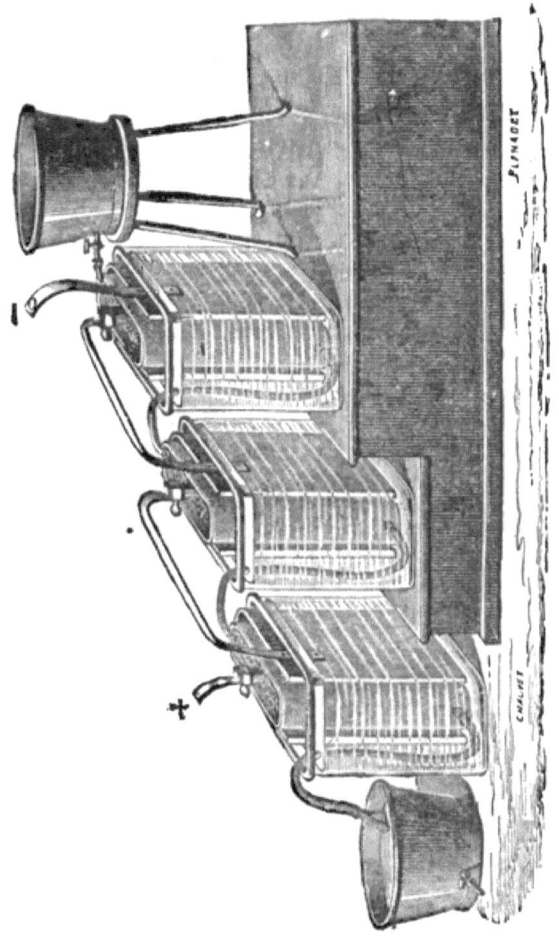

FIG. 51.

the bichromate-of-potash battery, by placing the cells one above the other and causing the liquid to pass successively in two or three cells. This disposition appears to us very cumbersome, and we prefer that of Camacho, represented by Fig. 51.

The jars are placed in gradation upon steps; the liquid falls from a special reservoir into the porous jar of the top cell, whence it is carried by means of a siphon into the succeeding one, and so on.

The negative electrode consists of a small bar of carbon and a considerable mass of crushed gas-retort carbon, which fills the porous jar; the enormous surface of this electrode renders polarization very slow.

We have recommended this disposition for all batteries subject to polarization, that is for those which have not an absolutely efficient depolarizing agent. In this battery it is very advantageously applied. Gaugain has published tables of comparison, which show the influence of crushed carbon placed around the principal plate.

DELAURIER'S BATTERY.

This is a modification of the bichromate-of-potash battery. It has in the outer jar a zinc electrode which surrounds the porous jar, in which is a carbon electrode. When the battery is charged, pure water is put with the zinc, and with the carbon a liquid composed of water, bichromate of potash, sulphate of soda, sulphate of iron, and sulphuric acid.

At first starting the resistance is considerable on account of the pure water in the outer jar; but a certain quantity of the compound liquid soon passes through the pores of the diaphragm and the resistance of the element diminishes about one half.

It is difficult to ascertain precisely what chemical

actions take place between these numerous compounds, either while the circuit is open or closed. However, Delaurier's battery is employed by many, and especially for depositing silver, for depositing nickel, gilding, etc.

The advantages of this battery are mostly attributed to the proportions of its different parts. The porous jar is comparatively large and contains two carbon plates instead of one, which diminishes the internal resistance.

PART III.
VARIOUS BATTERIES.

DRY PILES.

We will briefly describe those galvanic batteries called *dry piles*, in which the liquid is replaced by a slightly moistened or oily substance.

In reality, these batteries are not dry, and cease to work when they are thoroughly dried. In the beginning, cells were formed of two thin plates of copper and of zinc, having between them a sheet of paper saturated with oil or salt water and nearly dry.

In 1812 Zamboni suggested a more original disposition, which has been but slightly modified, and which is constructed as follows:

A sheet of paper is turned upon one side, and upon the other side is spread with a small brush a thin layer of peroxide of manganese thinned with milk or with gummy water, or even with common paste; this is left to dry, and then, placing a certain number of these sheets one upon another, small discs about $1\frac{1}{2}$ in. in diameter are cut out by means of a punch; the discs are then placed one upon another in regular order. To render this pile of discs solid, the latter are placed in a glass tube well varnished on the inside; they are pressed down as tightly as possible, in order to insure good contact. At both ends there are metallic plates which represent the poles of the battery. These metallic plates are a little larger in diameter than the discs of the battery and are pierced with several holes, through which silk cords are passed which hold the compressed column and even dispense with the glass tube of which we have spoken.

Great care must be taken, especially when the glass tube is not used, to exclude the air, which may be done by covering the pile with gum lac or with sulphur; otherwise losses would be sustained on account of the humidity of the air, and the effects of the battery would be greatly lessened.

Electric sparks can be produced with these batteries, as their electro-motive force is considerable on account of the large number of elements which enter into their composition.

The internal resistance of these batteries is enormous, as is easily understood. This explains why a most sensitive galvanometer is needed to detect any current.

It is easy to comprehend how dry piles recover but slowly their charge, after having been discharged, and how they do not receive their maximum charge in a damp atmosphere.

Zamboni's batteries do not last forever; it has been proved that after a few years they lose all their force. In their last stages of weakness, a portion of their energy may be restored by exposing them to an intense heat. It is probable that the chemical actions are thus accelerated.

The chemical reaction which takes place in each element is very simple; the peroxide of manganese is decomposed and the tin is oxidized at its expense. It is to be noted that the bioxide plays the part of both conducting electrode and of active substance. The paper acts as simple conductor, which property it owes to the humidity it contains. When this humidity disappears, the battery no longer produces any electricity, and it is only necessary to expose it to damp air in order that it may recover its electro-motive properties. Experiments show that dry piles possess all the properties of ordinary batteries.

Dry piles used to be employed to turn small apparatus based upon the attraction and repulsion of electrified bodies. Perpetual motion was thought to have been secured, but after a few years these apparatus stopped, proving once more the folly of such a research after the impossible. These scientific playthings have now gone out of fashion; no more are constructed, but they are found described and illustrated in nearly all works upon physics.

To-day dry piles are only used in connection with electroscopes, notably with that of Bohnenberger, the description of which is found in most works treating of the subject.

IDENTICAL ELECTRODE BATTERIES.

Hitherto in batteries we have always met with two distinct electrodes immersed in one or two liquids.

We desire to show that batteries may be composed which contain identical electrodes, provided these electrodes be immersed in two different liquids. The direction of the current will be determined because one of the liquids will attack its electrode more actively and the other will attack its electrode less actively or not at all.

We will give a very simple example. In a jar put a saturated solution of sulphate of copper, and on top of this some acidulated water, or salt water, or pure water. Now if a copper plate be placed on end in this liquid, it will be attacked at the top and become charged with copper at the bottom. It is plain that we have there an element of copper, acidulated water, sulphate of copper, and copper.

This experiment furnishes the explanation of how a drop of sulphate of copper thrown upon copper will produce an attack, the same as a drop of nitrate of silver upon a silver plate. It happens that, by difference of specific gravity, the more saturated solution falls to the bottom and the less charged solution rises to the top; consequently a current is produced which transfers the metal from one point to another. In order to have no action at all, the composition of the liquid ought to be the same throughout.

We thought it better to note these observations to show the reader that various phenomena not apparently electric can in reality only be explained by electro-chemistry.

UNATTACKED ELECTRODES IN BATTERIES.

In all the batteries hitherto described we have found a metallic generating electrode. This electrode is dissolved in the liquid in which it is immersed, and this action produces a current. But the actions between liquids can also produce electricity.

If in a vessel divided into two compartments by a porous partition acid is put in one division and an alkaline solution in the other, there will be a reciprocal reaction in the pores of the diaphragm and the formation of a salt; if two platinum plates be immersed in the liquids a current will be made manifest in a wire conductor uniting the two plates. The positive pole of this element will correspond to the acid. Nothing would be changed if, instead of two platinum plates immersed in the two compartments, there were one of platinum and one of gold. Neither of them are attacked, both are conducting electrodes, and there is in reality no generating electrode.

We will not insist upon these voltaic contrivances which have as yet received no application. It should be noticed, however, that they generally polarize like ordinary batteries. A single one has been invented which does not polarize, and which should be noted in the enumeration that we have undertaken.

BECQUEREL'S OXYGEN-GAS BATTERY.

This apparatus is composed as follows: A glass tube closed at the bottom by a porous partition is placed in a flask containing nitric acid; potash is poured in the glass tube, and in each liquid is placed a bar of platinum to which a conductor is attached. When the circuit is closed by bringing the two connections in contact with each other, a lively action takes place, the acid and the base combine. But this simple reaction does not take place alone; the current thus produced decomposes the surrounding water, the hydrogen goes to the nitric acid, which it reduces, and is not given off upon the electrode; consequently there is no polarization. The oxygen goes to the potash, remaining free, and surrounds the platinum plate.

This battery is remarkable as being the only one in which oxygen is evolved, the contrary from all other batteries in which hydrogen is evolved (except, of course, totally depolarized batteries).

It moreover possesses a great historical interest, because it is the first battery furnishing a constant current that was constructed; it is in this element that were first employed two liquids and a porous diaphragm.

COKE-CONSUMING BATTERY.

Throughout this work we have shown that the combustion or dissolving of zinc is the principal and almost only means employed to produce voltaic electricity. Thus, each equivalent of electricity costs, at the minimum, one equivalent of zinc plus an equivalent of one or two other substances, and it is for that reason that electricity is so expensive.

It might be reasonably inquired whether the combustion of common coal could not be utilized for the production of electric currents.

Magneto-electrical machines, which furnish a continuous current, and of which the Gramme machine is the model, present a solution of this problem. When a Gramme machine is put into motion by a steam motor, there is seen a transformation into electricity of the heat produced by the combustion of the coal in the furnace of the motor. That is an indirect solution, for the heat is first transformed into motive force, which is in turn transformed into electricity.; but it is a very good solution, and is to-day confirmed by practice. If the electricity produced by Gramme machines cost but little, it is because it is produced by the combustion of coal, which is as yet the most advantageous source of energy discovered—an energy which presents itself in the form of heat, of chemical energy, of movement, or of electricity, and which may be transformed into one or the other of these four powers.

It is very probable that a direct transformation into

electricity of the heat produced by the combustion of coal may be obtained. Mr. Jablochkoff has already invented a battery cell which fulfils the above conditions.

The liquid of this cell is melted nitrate of potash or nitrate of soda; one electrode is of coke, and the other of platinum, or even of cast-iron. The coke is burned at the expense of the oxygen of the nitrate, and produces torrents of carbonic acid; the cast-iron remains unattacked.

The coke is therefore the positive electrode, and the cast-iron the negative. It is the contrary of that which would take place in a battery with an ordinary liquid, acid, or salt dissolved in water.

The nitrate should be previously melted, but as soon as the action begins the salt remains liquid on account of the great heat produced by the combinations which take place; and if the element be left to itself it suffices, to put it in action, to bring the end of the coke to a glowing red heat, and then to press it against the surface of the salt; the chemical action begins immediately, and by the heat it produces the melting of the nitrate, and soon the element is reproduced.

It might be found that such a battery cell presents nothing practical in its actual form, and we do not hesitate to express that opinion, but we believe that it points out a new way in which much progress might be made if the attention of physicists were turned in that direction. Volta's battery itself when invented was a purely scientific novelty, and it was far from being regarded as an object of any practical utility.

Mr. Jablochkoff's experiment will doubtless call forth many commentaries. This is not the place for any remarks upon the subject. We only wished, in mentioning this battery cell, to show our readers that beyond the already explored horizons there remain other worlds to discover and virgin lands to cultivate.

GAS BATTERIES.

We have explained how a voltameter, in which oxygen and hydrogen were evolved, could become a source of electricity. To show this, it is only necessary to attach the wires of a galvanometer to the terminals of a voltameter thus charged.

If the electrodes of the voltameter are immersed partly in the acidulated water and partly in the previously evolved gases, the voltameter may furnish a current during a considerable length of time; the oxygen and hydrogen recombine through the liquid, and the current resulting from this combination lasts as long as there is any gas in contact with one or the other of the electrodes.

Under these circumstances, a voltameter becomes in reality a gas battery. When the gases are consumed they can be replaced, and the action of the battery prolonged, just as acids or salts are renewed in ordinary batteries.

Grove, to whom is due the invention of these interesting contrivances, has given to them many and various forms. He has also substituted for the oxygen and the hydrogen other gases, such as chlorine, protoxide of carbon, bioxide of nitrogen, olefiant gas, and has found batteries analogous to the first one. He has discovered that certain gases when taken together produce no electric current; viz., nitrogen and oxygen, nitrogen and hydrogen, protoxide of nitrogen and oxygen.

A certain number of cells, such as we have described, may be joined in intensity, and it will be seen that they possess the same general properties as ordinary batteries.

Gaugain has established the fact that gas batteries do not act except by dissolved gases; so that their force becomes less as the solution weakens, just as in sulphate-of-mercury and sulphate-of-lead cells.

It is moreover probable that, in addition to this special weakening, gas batteries are liable to polarize like the others, especially when there are several elements in the circuit.

Batteries in which only one of the active bodies is gaseous, while the other is liquid, have also been discovered.

Thus Grove has been able to obtain a current by causing oxygen to act upon sulphate of protoxide of iron, which is subject to oxidation and is transformed into sulphate of peroxide, or by causing hydrogen to act upon nitric acid, which decomposes and gives off oxygen.

Ed. Becquerel caused hydrogen to act upon chloride of gold in the presence of platinum.

Two or single gas batteries can receive no application; the great interest they possess is simply theoretical.

SECONDARY BATTERIES.

We have already said that a voltameter submitted to the influence of an electric current for a moment becomes capable of furnishing a current contrary to the exciting current. This capital fact has enabled us to show, under one of its plainest forms, the phenomenon of the polarization of electrodes.

The current thus furnished by the voltameter is a *sec-*

ondary current, and the apparatus becomes a *secondary element*. The current may be said to have been furnished by the battery, and returned by the secondary element.

The study of this question dates from the beginning of the present century—Gautherot in 1801, and soon after Ritter called attention to it. Ritter has shown that secondary elements may be produced under other forms than that of the voltameter; two electrodes of platinum, of carbon, of copper, or indeed of more easily oxidized metals, placed in a conducting liquid, suffice to constitute a secondary element.

A column battery formed of a succession of discs of copper and cloth moistened with sulphate of potash constitutes a secondary battery.

Pursuing the general idea with which this work has been written, we can, in conclusion, do no better than to study in detail secondary batteries, which present an application of the polarization of electrodes.

To make the subject clear, let us consider a secondary battery formed of two platinum plates immersed in acidulated water.

The most feeble current suffices to polarize these electrodes, as can be shown by carefully observing the secondary current. It is not necessary that the exciting current should have a tension sufficiently great to decompose water; one of Daniell's elements, or even a more feeble one, can polarize the secondary element.

When the cell is polarized by a more energetic current, the secondary current also increases in energy, but it is clear that the electro-motive force of the second can never be superior to that of the first.

If in a circuit are placed an ordinary cell, a galvanometer, and a secondary cell, the following facts are observed: at first the current of the polarizing cell passes with great energy, as the galvanometer shows, then it gradually diminishes on account of the increasing polarization of the secondary cell.

If the polarizing current is not sufficient to decompose water—if, for instance, it is furnished by one of Daniell's cells—it happens that the secondary current counterbal-

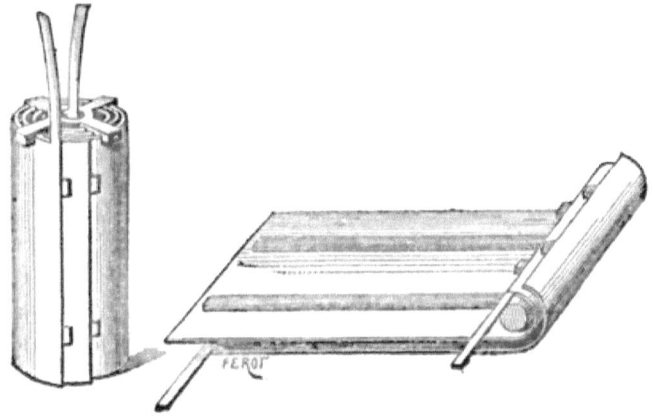

Fig. 52.

ances the principal current, as shown by the galvanometer, which marks no deflection.

If the principal current is sufficient to decompose water, the above equality is never reached, and the freeing of gases corresponds indefinitely to a circulation of the current made manifest by the galvanometer.

If, instead of a single secondary element, several are placed in the circuit, polarization will be divided between them; each one taken separately will furnish a secondary current after being a moment under the influence of the principal current, but the sum of the electro-motive forces of these secondary currents can never become superior to that of the polarizing current.

But if these secondary cells be charged separately and then joined in intensity, the total current might have a

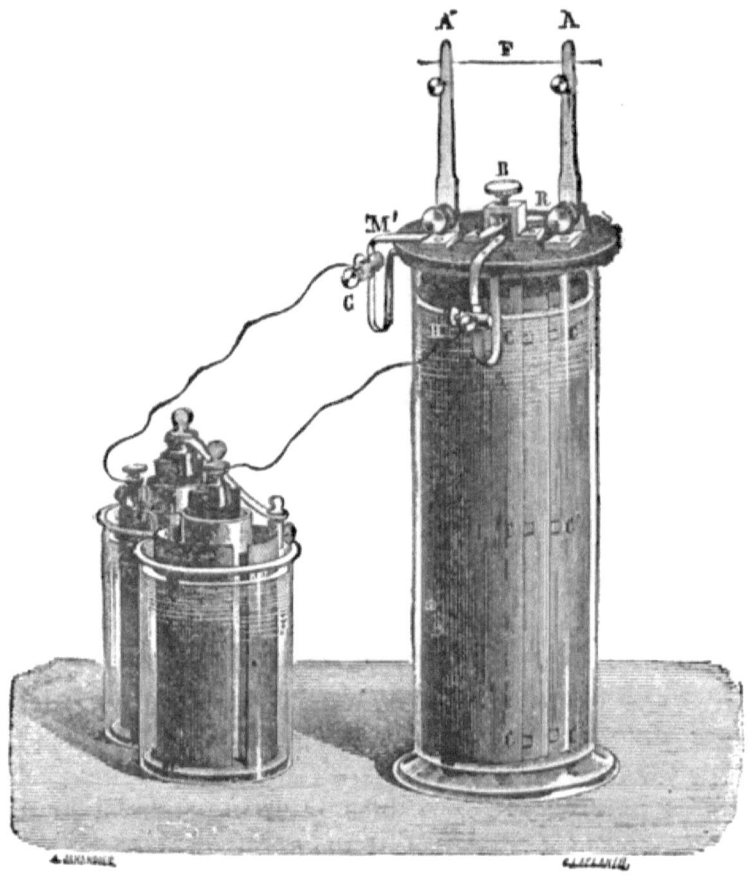

Fig. 53.

considerable energy and be superior to that of the principal current by which they have been successively excited.

All this will be made clearer by the following detailed study of secondary cells having electrodes of lead.

As early as 1859 Mr. Planté showed that lead was the most favorable metal for use in secondary batteries, and he has since that time accumulated many proofs of this superiority. Figs. 52 and 53 show the element as constructed to-day. In a tall vessel made of glass, of rubber, or of ebonite are placed two sheets of lead rolled together parallel to each other, and kept apart by two strips of rubber rolled with them; these two sheets are immersed in a solution containing one tenth of sulphuric acid. The vessel is closed by a sealed stopper in which there is a hole through which the liquid is introduced and extracted, and through which the gases evolved during the charging may pass off. The apparatus is capped by an ebonite cover furnished with two clamp screws which communicate with the two electrodes; there are also two clamps, which hold metallic wires to be heated and melted by the secondary current.

To charge this secondary element to its maximum, two of Bunsen's cells or three of Daniell's must be used. During the charging, one of the electrodes becomes oxidized, a brownish layer of peroxide of lead is soon seen, and the metallic aspect completely disappears; the other electrode only changes in appearance, its surface becoming covered with a grayish matter.

When it is charged to its maximum—that is, when oxygen begins to free itself from the brown electrode—it is well to separate the secondary cell from the active battery, as the polarizing current is no longer useful and is wasted.

The secondary element thus charged and left to itself can preserve a part of its charge several days, and at the end of a week it is still far from being exhausted.

The secondary cell when charged to its maximum has an electro-motive force equal to one and a half that of Bunsen's cell; it can bring to a glowing red heat a platinum wire large or small according to the dimensions of the cell, or, better, according to the size of its electrodes. It can be easily understood, indeed, that the quantity of electricity furnished by the apparatus is in proportion to the extent of surface of the lead submitted to the action of the polarizing current and covered with an active electro-chemical deposit.

It should be noted that the peculiar form of the electrodes offers a large surface and a small resistance under a small volume; so that one of Planté's secondary cells is equal to an active or ordinary cell of extraordinary dimensions; the small model has a surface of eight square decimetres, the large one a surface of four square decimetres.

The current furnished by a secondary element can produce chemical decompositions, act upon an electromagnet, etc.; but if its intensity be measured in one way or another, with a galvanometer for instance, it is seen to diminish from the maximum of which we have spoken above. This decrease is very slow if the circuit offers a great resistance, and if, as a consequence, there is a very small flow of electricity; it is on the contrary very rapid if the circuit offers but a slight resistance, because the electricity flows in a large quantity.

A very curious and interesting fact is noticed during the discharge of the cell; it is apparently completely discharged, but if the circuit be left open several minutes, it has been ascertained that it recovers a certain energy and that it can still furnish a certain quantity of electricity.

The battery thus delivered of its first *residue* and left to itself for some time will furnish a second residue, less, of course, than the first. And this is not the last one, for several more can be obtained. Mr. Planté has very clearly explained this peculiarity. The secondary element, when it becomes active, discharges itself and at the same time polarizes, as all single-liquid batteries. This polarization attains in a certain time a force almost equal to that of the already weakened secondary element, and the action ceases or is reduced to very little. If the battery then be left to rest, it depolarizes as do all single-liquid batteries polarized by their own action. As soon as the battery is depolarized it is again ready to furnish a current, but during this new discharge it again polarizes, and so on.

If we consider the secondary cell as completely or almost completely discharged, it may be recharged with two of Bunsen's elements, as in the first instance; but it is well to note that the more immediately after the discharge the new charge be given, the more rapidly it may be given.

Moreover, the greater number of times a secondary element is charged and discharged, the better it is. In the beginning, when it is nearly new, there is an advantage in polarizing the electrodes, first in one direction and then in the other, and in reversing several times the direction of the charge; but when the element is formed, great care must be taken to always charge it in the same direction. If this precaution be neglected it will take a much longer time to charge it, for the oxide of lead, which may still remain upon one of the electrodes, must be reduced and the previously negative plate oxidized. But after this operation the secondary cell will have recovered all its qualities: it may indeed be said to have gained some.

Fig. 54 shows a peculiar form given to the secondary element by Mr. Planté, and which he has called *Saturn's tinder-box*. At the top are seen two clamps which hold a platinum wire stretched between them; each time that, by pressing with the finger, the two springs at the bottom are brought into contact the battery sends a current through the platinum wire, which is thereby brought to a glowing red heat, whence follows an almost instantaneous lighting of the candle. With one of these contrivances the candle may be lighted one hundred times, and it is only after these frequent lightings that it has to be recharged with three of Daniell's cells. That is a means of obtaining fire and a very economical means too, for the secondary element spends nothing and the charging battery consumes but a few grammes of sulphate of copper for a prolonged working of the tinder-box.

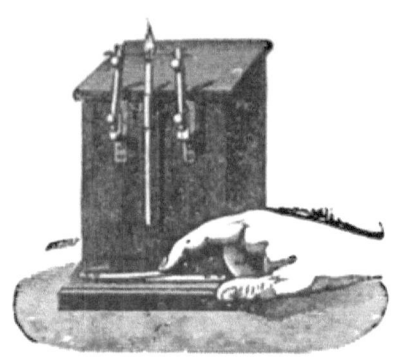

Fig. 54.

This same apparatus can be used to touch off mines either in civil or military service; the experiment shows that with fine platinum-wire fuses ($\frac{1}{300}$ of an inch) combustion may be obtained through a copper wire 1000 yards long.

With a contrivance of this kind surgeons may cauterize a wound, and it has frequently been applied in that way. A secondary element is much more easily transported into a hospital or to the house of a sick person than active cells which it may replace.

Finally, secondary cells can be joined in quantity or in intensity and constitute batteries capable of producing all the effects of the most powerful ordinary batteries. Fig. 55 represents the secondary battery as disposed by Mr. Planté.

The number and dimensions of the cells can be varied according to the tension and quantity desired. Here there

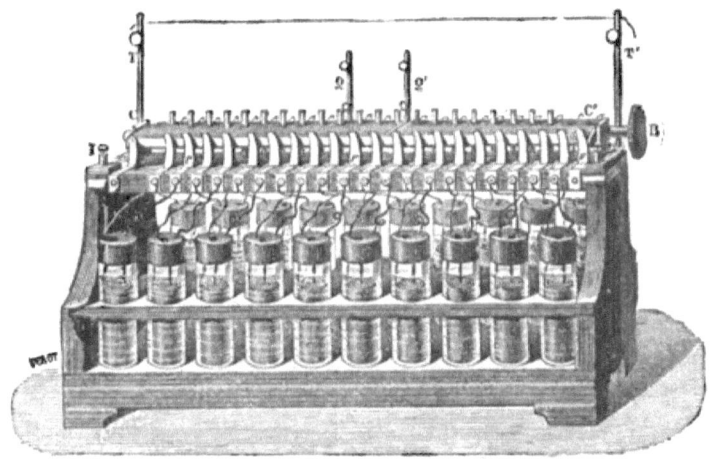

Fig. 55.

are twenty elements arranged in two rows. At the top there is a very conveniently disposed commutator, which, in one position, joins the cells in quantity; in another position, at right angles with the first, it joins them in intensity. In the first position all the outer electrodes are joined to one metallic strip, and all the inner electrodes to another metallic strip, so that the whole arrangement represents a single cell with a large surface. It is in this condition that the charge is made; two of Bunsen's cells are sufficient, and they complete the charge in a longer or shorter time, according to the dimensions of the cells and to the extent of the surface of the lead to be polarized. In the second position the outer electrode of each cell is put into communication with the inner one of the following cell, and the apparatus becomes a real battery of twenty cells. It is in this condition that the battery is discharged, and it is equal, at first starting, to 30 of Bunsen's very large cells.

As the battery is being discharged the tension diminishes, as we explained when speaking of the single secondary element. If it takes one minute to charge the battery of secondary cells in quantity, it cannot be expected that the discharge of the cells in intensity will furnish the same effects as 30 of Bunsen's of the same size during a longer period than four seconds, for the apparatus furnishes no electricity and can only transform that which has been given to it. Mr. Planté has made some exact experiments in this direction, and has found that in this transformation about one tenth is lost, or, in other words, the machine returns nine tenths of that which was given to it.

It is clearly seen that the secondary battery can only produce effects of very short duration, but in most cases this is all that is neccesary.

If, for instance, a large number of mines are to be simultaneously exploded by means of fuses of fine wire, it may be done by placing all the fuses in divided circuits and by causing the current of the secondary battery to pass through them all at once. This manner of proceding is very economical, and it is certainly less laborious and costly to mount two of Bunsen's cells and to charge a secondary battery than to charge 20 or 30 of Bunsen's elements, especially when the battery is only worked a few seconds and only four or five times during a day.

I. ELECTRIC CONDUCTING POWER OF SOLIDS.

(*Ed. Becquerel.—Annales de Chimie et de Physique*, 1846.)

Substances.	Metal (hard drawn).	Metal (annealed).
Silver, pure (reduced from the chloride)....	93.448	100.000
Copper, pure electro-chemical................	89.048	91.439
Gold, pure..................................	64.385	65.458
Cadmium.....................................	24.574	......
Zinc...	24.164	......
Tin...	13.656	......
Palladium....................................	13.977	......
Iron...	12.124	12.246
Lead...	8.245	......
Platinum.....................................	8.042	8.147
Mercury at 14° centig........................	1.8017	......

Substances.	Conductivity at 0° Centig.	Conductivity at 100° Centig.	Coefficient for 1° Centig.
Silver, annealed............................	100.	71.316	0.004022
Copper......................................	91.517	64.919	0.004097
Gold..	64.960	48.489	0.003397
Cadmium.....................................	24.579	17.506	0.004040
Zinc...	24.063	17.596	0.003675
Tin..	14.014	8.657	0.006188
Iron, annealed..............................	12.350	8.387	0.004726
Lead...	8.277	5.761	0.004349
Platinum, annealed..........................	7.933	6.688	0.001861
Mercury, distilled..........................	1.7387	1.5749	0.001040

II. SPECIFIC RESISTANCES DETERMINED BY MATTHIESEN.

(*Table taken from Fleeming Jenkin.*)

NAMES OF METALS.	Resistance of one cubic centimetre to conduction between opposed faces.	Resistance of a wire one metre long and one millimetre in diameter.	Resistance of a wire one metre long weighing one gramme.	Resistance of a wire one foot long, $\frac{1}{1000}$ of an inch in diameter.	Resistance of a wire one foot long weighing one grain.
	Microhms.	Ohms.	Ohms.	Ohms.	Ohms.
Silver, annealed.................	1.521	0.01937	0.1544	9.151	.2214
Silver, hard drawn............	1.652	0.02103	0.1680	9.936	.2415
Copper, annealed.............	1.616	0.02057	0.1440	9.718	.2064
Copper, hard drawn...........	1.652	0.02104	0.1469	9.940	.2106
Gold, annealed.................	2.081	0.02650	0.4060	12.52	.5849
Gold, hard drawn.............	2.118	0.02697	0.4150	12.74	.5950
Aluminium, annealed	2.945	0.03751	0.0757	17.72	.1085
Zinc, pressed..................	5.689	0.07244	0.4067	34.22	.5831
Platinum, annealed	9.158	0.1166	1.96	55.09	2.810
Iron, annealed.................	9.825	0.1251	0.7654	59.10	1.097
Nickel, annealed	12.60	0.1604	1.071	75.78	1.535
Tin, pressed...................	13.36	0.1701	0.9738	80.36	1.396
Lead, pressed..................	19.85	0.2526	2.257	119.39	3.236
Antimony, pressed............	35.90	0.4571	2.411	216	3.456
Bismuth, pressed	132.7	1.689	13.03	798	18.64
Mercury, liquid...............	99.74	1.2247	13.06	578.6	18.72
Platinum silver...............	24.66	0.3140	2.959	148.35	4.243
German silver, hard or annealed...................	21.17	0.2695	1.85	127.82	2.652
Gold-silver alloy, hard or annealed: two parts gold, one part silver............	10.99	0.1399	1.668	66.10	2.391

	Resistance of one cubic centimetre between opposed faces, expressed in microhms.	Temperature, Centigrade.
Graphite specimen, No. 1...............	2,300	22°
" " No. 2...............	3,780	22°
" " No. 3...............	41,800	22°
Gas-retort carbon..................	4,280	25°
Carbon in Bunsen's battery............	67,200	26.2°
Tellurium.......................	212,500	19.6
Red phosphorus..................	132 ohms	20°

III. CONDUCTIVITY OF LIQUIDS.

(*Ed. Becquerel.—Annales de Chimie et de Physique*, June 1846.)

Substances.	Specific Weight.	Temperature, Centigrade.	Conductibility.	Coefficient of the increase of conductibility for 1° centigr.	Observations.
Silver............	....	0°	100,000,000.00		
Sulphate of copper, saturated....................	1.1707	9.25	5.42	0.0286	
Sulphate of copper, diluted to half.............	1.1707	9.25	3.47	....	
Sulphate of copper, diluted to quarter..........	1.1707	9.25	2.08	....	
Chloride of sodium, saturated....................	1.1707	13.40	31.52	....	
Chloride of sodium, diluted to half.............	1.1707	13.40	23.08	....	
Chloride of sodium, diluted to third............	1.1707	13.40	17.48	....	
Chloride of sodium, diluted to quarter..........	1.1707	13.40	13.58	....	
Bichloride of copper, saturated and diluted with five times its bulk of water	1.1707	13.40	10.35	....	
Nitrate of copper, saturated	1.6008	13.00	8.995	....	
Nitrate of copper, diluted to ⅔................	....	13.00	16.208	....	
Nitrate of copper, diluted to half..............	....	13.00	17.073	....	Maximum.
Nitrate of copper, diluted to quarter....... .	....	13.00	13.412	....	
Sulphate of zinc, saturated	1.4410	14.40	5.77	0.0223	
Sulphate of zinc, diluted to half.................	1.4410	14.40	7.13	....	Maximum.
Sulphate of zinc, diluted to quarter..............	1.4410	14.40	5.43	....	
Iodide of potassium, 30 gr.; water, 250 gr.......	1.4410	12.50	11.20	....	
Monohydrate sulphuric, 20 gr.; water, 220 gr........	1.4410	19.00	88.68	....	
Nitric acid, commercial (sp. w. 1.31).............	1.4410	13.10	93.77	0.0263	
Protochloride of antimony, 30 gr.; water, 120 gr.; and hydrochloric acid, 100 gr...................	1.4410	15.00	112.01	....	

This table shows the maximum conductibilities of the solutions of nitrate of copper and sulphate of zinc, but not that of chloride of sodium. The maximum conductibility of this latter is found in a mixture of 24.4 parts of chloride for 100 of water.

IV. LIQUID RESISTANCES.

(*Table taken from Fleeming Jenkin. Calculated by Becker.*)

SULPHATE OF COPPER.

Percentage of salt in solution.	TEMPERATURE, CENTIGRADE.							OBSERVATIONS.
	14°	16°	18°	20°	24°	28°	30°	
8	45.7	43.7	41.9	40.2	37.1	34.2	32.9	Resistance of a
12	36.3	34.9	33.5	32.2	29.9	27.9	27.0	cubic centimetre
16	31.2	30.0	28.9	27.9	26.1	24.6	24.0	expressed in
20	28.5	27.5	26.5	25.6	24.1	22.7	22.2	ohms.
24	26.9	25.9	24.8	23.9	22.2	20.7	20.0	
28	24.7	23.4	22.1	21.0	18.8	16.9	16.0	

SULPHURIC ACID—DILUTED.

SPECIFIC GRAVITY.	0°	4°	8°	12°	16°	20°	24°	28°	
1.10	1.37	1.17	1.04	.925	.845	.786	.737	.709	Resistance of one
1.20	1.33	1.11	.926	.792	.666	.567	.486	.411	cubic centi-
1.25	1.31	1.09	.896	.743	.624	.509	.434	.358	metre to con-
1.30	1.36	1.13	.94	.79	.662	.561	.472	.394	duction be-
1.40	1.60	1.47	1.30	1.16	1.05	.964	.896	.839	tween opposed
1.50	2.74	2.41	2.13	1.89	1.72	1.61	1.52	1.43	faces expressed
1.60	4.82	4.16	3.62	3.11	2.75	2.46	2.21	2.02	in ohms.
1.70	9.41	7.67	6.25	5.12	4.23	3.57	3.07	2.71	

SULPHATE OF ZINC.

	10°	12°	14°	16°	18°	20°	22°	24°	
96 grammes in 100 c.c. of solution........	22.7	21.4	20.2	19.2	18.1	17.1	16.3	15.6	Resistance of one cubic centimetre.

	10°	12°	14°	16°	18°	20°	22°	24°	
The same solution with an equal volume of water......	......	......	21.1	20.3	19.5	18.8	18.1	17.3	Expressed in ohms.

	2°	4°	8°	12°	16°	20°	24°	28°	
Nitric acid (sp. w. 1.36)................	1.94	1.83	1.65	1.50	1.39	1.30	1.22	1.18	Resistance of one cubic centimetre in ohms.

V. DILUTE SULPHURIC ACID.

(*Bineau's Table.*)

Degrees upon the Hydrometer.	Specific Weight.	TEMPERATURE = 0° CENTIG.		TEMPERATURE = 15° CENTIG.	
		Monohydrate acid for 100 of the mixture.	Anhydride acid for 100 of the mixture.	Monohydrate acid for 100 of the mixture.	Anhydride acid for 100 of the mixture.
5.0	1.060	5.1	4.2	5.4	4.5
10.0	1.075	10.3	8.4	10.9	8.9
15.0	1.116	15.5	12.7	16.3	13.3
20.0	1.161	21.2	17.3	22.4	18.3
25.0	1.209	27.2	22.2	28.3	23.1
30.0	1.262	33.6	27.4	34.8	28.4
33.0	1.296	37.6	30.7	38.9	31.8
35.0	1.320	40.4	33.0	41.6	34.0
36.0	1.332	41.7	34.1	43.0	35.1
37.0	1.345	43.1	35.2	44.3	39.2
38.0	1.357	44.5	36.3	45.5	32.2
39.0	1.370	45.9	37.5	46.0	38.3
40.0	1.383	47.3	38.6	48.4	39.5
41.0	1.397	48.7	39.7	49.9	40.7
42.0	1.410	50.0	40.8	51.2	41.8
43.0	1.424	51.4	41.9	52.5	42.9
44.0	1.438	52.8	43.1	54.0	44.1
45.0	1.453	54.3	44.3	55.4	45.2
46.0	1.468	55.7	45.5	56.9	46.4
47.0	1.483	57.1	46.6	58.2	47.5
48.0	1.498	58.5	47.8	59.6	48.7
49.0	1.514	60.0	49.0	61.1	50.0
50.0	1.530	61.4	50.1	62.6	51.1
51.0	1.546	62.9	51.3	63.9	52.2
52.0	1.563	64.4	52.6	65.4	53.4
53.0	1.580	65.9	53.8	66.9	54.6
54.0	1.597	67.4	55.0	68.4	55.8
55.0	1.615	68.9	56.2	70.0	57.1
56.0	1.634	70.5	57.5	71.6	58.4
57.0	1.652	72.1	58.8	73.2	59.7
58.0	1.671	73.6	60.1	74.7	61.0
59.0	1.691	75.2	61.4	76.3	62.3
60.0	1.711	76.9	62.8	78.0	63.6
61.0	1.732	78.6	64.2	79.8	65.1
62.0	1.753	80.4	65.7	81.7	66.7
63.0	1.774	82.4	67.2	83.9	68.5
64.0	1.796	84.6	69.0	86.3	70.4
65.0	1.819	87.4	71.3	89.5	73.0
65.5	1.830	89.1	71.2	91.8	74.9
65.8	1.837	90.4	73.8	94.5	77.1
66.0	1.842	91.3	74.5	100.0	81.6
66.2	1.846	92.5	75.5	....	....
66.4	1.852	95.0	77.5	....	....
66.6	1.857	100.0	81.6	....	....

VI. RESISTANCE OF DIFFERENT LIQUIDS.

Dilute Sulphuric Acid after Saweljev. (Extract from Wiedemann.)				Chloride of Sodium.		Nitrate of Potash.	
Specific Weight.	Parts by weight of SO_3HO for 100 of water.	Temperature, Centigrade.	Resistance.	Parts by weight for 100 grams of water.	Resistance at 18° centigrade.	Parts by weight for 100 grams of water.	Resistance at 18° centigrade.
1.003	0.5	16.1	16.01	23.8758	0.59852	18.9167	0.83271
1.018	2.2	15.2	5.47	24.4083	0.57982 *	13.7647	1.10626
1.053	7.9	13.7	1.884	20.9787	0.63840	10.4840	1.35099
1.080	12.0	12.8	1.368	17.0174	0.71109	6.6079	1.94955
1.147	20.8	13.6	0.960	10.4525	1.03934	3.3964	3.32633
1.190	26.4	13.0	0.871	6.0037	1.55599	1.5432	6.38318
1.215	29.6	12.3	0.830	3.6880	2.46192		
1.225	30.9	13.6	0.862	1.7177	5.56371		
1.252	34.3	13.5	0.874				
1.277	37.3	....	0.930	* Minimum.			
1.348	45.4	17.9	0.973				
1.393	50.5	14.5	1.086				
1.492	60.6	13.8	1.549				
1.638	73.7	14.3	2.786				
1.726	81.2	16.3	4.337				
1.827	92.7	14.3	5.320				

The above figures show the maximum conductibility of the mixture to be that of 29 to 30 parts of the monohydrate acid for 100 of water; a little different from that of the preceding table.

The above figures are taken from a memoir of Schmidt; *Annales de Poggendorff.* See Wiedemann, vol. i. p. 324.
It is seen that the maximum conductibility or the minimum resistance of the sea-salt solution corresponds to 24.4 for 100 of water.
The figures correspond to the Jacobi's standard of resistance, and must be multiplied by 508×10^7 to be brought to electro-magnetic absolute measurements.

Experiments of Horsford, 1847. (*See Wiedemann.*)

Chloride of potassium, 27.6 grammes in 500 grammes of water....	577,100
" " diluted to half............................	1,103,700
" " quarter.................................	2,006,500
Chloride of sodium, 27.6 grammes in 500 grammes of water.......	577,100
" " diluted to half............................	1,488,200
Chloride of calcium, dissolved (sp. w., 1.04)......................	672,560
Chloride of magnesium..	672,560
Chloride of zinc...	1,092,500

Experiments of Wiedemann (1856) *from* 18° *to* 20° *Centigrade.*

SULPHATE-OF-COPPER SOLUTION.

31.17 grammes in one litre of water...........................	7,803,000
62.34 " " "	4,202,000
77.92 " " "	3,514,000
93.51 " " "	3,178,000
124.68 " " "	2,507,000
155.85 " " "	2,181,000
187.02 " " "	1,930,000

VII. ELECTRO-MOTIVE FORCES. (J. REGNAULD, 1854.)

						Volts
Wheatstone	Zinc	Sulphate of zinc	Sulphate of cadmium	Cadmium	1	0.0046
	Copper	Hydrate of potash	Sulphate of copper	Copper	55	0.500
	Amalgam of zinc-zinc 1, mercury 15				90	0.4914
Daniell	Zinc	Sulphate of copper	Sulphate of copper	Copper	153	0.885
	Amalgamated zinc	Sulphate of zinc	Sulphate of copper	Copper	165	0.901
Grove	Amalgamated zinc	SO₃HO + 10 water by wght	Sulphate of copper	Copper	173	0.945
Joule	Amalgamated zinc	SO₃HO + 10 water by wght	Nitric acid	Platinum	210	1.692
	Amalgamated zinc	Sulphate of soda	Nitric acid	Platinum	211	1.698
	Amalgam of potassium—mercury 150 and potassium 1	Sea-salt	Nitric acid	Platinum	224	1.769
Joule	Amalgamated zinc	Chloride of platinum	Chloride of platinum	Platinum	417	2.276
		Hydrate of potash	SO₃HO + 10 HO	Deposited peroxide of lead		
Regnauld	Zinc	Chloride of zinc	Chloride of cadmium	Cadmium	466	2.544
	Zinc	Bromide of zinc	Bromide of cadmium	Cadmium	42	0.229
	Zinc	Iodide of zinc	Iodide of cadmium	Cadmium	42	0.229
	Zinc	Nitrate of zinc	Nitrate of cadmium	Cadmium	45	0.246
	Amalgamated zinc	Sulphate of zinc	Sulphate of cadmium	Cadmium	42	0.229
	Amalgamated zinc	SO₃HO + 5 HO	Sulphate of cadmium	Cadmium	58	0.316
	Zinc	Nitrate of zinc	Chloride of cobalt	Cobalt	59	0.322
	Zinc	Sulphate of zinc	Nitrate of cobalt	Cobalt	114	0.622
	Zinc	Chloride of zinc	Sulphate of nickel	Nickel	94	0.513
	Zinc	Nitrate of zinc	Chloride of nickel	Nickel	127	0.693
	Zinc	Sulphate of zinc	Nitrate of nickel	Nickel	109	0.595
	Zinc	Nitrate of zinc	Sulphate of copper	Copper	131	0.715
	Zinc	Acetate of zinc	Nitrate of copper	Copper	175	0.955
	Zinc	Formiate of zinc	Acetate of copper	Copper	160	0.873
	Zinc	Chloride of zinc	Chloride of copper	Copper	175	0.955
	Amalgamated zinc	SO₃HO + 10 water by wght	Sulphate of copper	Copper	175	0.955
	Amalgamated zinc	SO₃HO + 10 water by wght	Nitric acid 1 vol, water 5 vol	Copper	179	0.978
	Zinc	Sulphate of zinc	Sulphate of alumina	Platinum	310	1.692
	Zinc	Dilute sulphuric acid	Sulphate of protoxide of mercury	Aluminium	35	0.191
				Carbon	220	1.200

VIII. ELECTRO-MOTIVE FORCES.

(*Poggendorff*, 1845.—*Extract from Wiedemann*.)

SINGLE-LIQUID BATTERIES.

Zinc..................		Daniell =	1.000
Tin...................		Tin	0.409
Zinc..................		Copper.....	0.410
Iron..................	Sulphuric acid (sp. w. = 1.838) diluted with 49 times its weight of water......................	Copper.....	0.824
Zinc..................		Copper.....	0.417
Zinc..................		Silver	1.053
Cadmium............		Cadmium...	0.339
Amalgamated zinc..		Iron........	0.191
Amalgamated zinc..		Iron........	0.537
		Tin..........	0.531
Amalgamated zinc..	Nitric acid (sp. w. = 1.22) diluted with 9 times its w'ght of water	Copper.....	0.682
Amalgamated zinc..		Platinum ..	1.495
Amalgamated zinc..		Copper.....	0.788
Amalgamated zinc..	Hydrochloric acid (sp. w.=1.113), diluted with 9 times its weight of water	Platinum...	1.537
Copper...............		Platinum...	0.771
Silver		Platinum...	0.620
Zinc..................		Iron	1.008
Zinc..................	Potash in 6 times its weight of water	Silver	1.198
Zinc..................		Platinum...	1.257
Zinc..................		Antimony..	0.541
Zinc..................	Carbonate of potash............	Iron........	0.832
Zinc..................	Carbonate of potash, concentrated	Copper.....	0.909
Zinc..................		Platinum...	1.078
Iron..................		Copper.....	0.072
Zinc..................	Chloride of potassium...........	Iron........	0.476
Zinc..................	Chloride of potassium, concentrated	Copper.....	0.743
Zinc..................		Platinum...	1.346
Iron..................		Copper.....	0.260

TWO-LIQUID BATTERIES.

	Iron.	$SO_3HO + 49HO$ by weight............	Sulphate of copper.	Copper.....	0.461
	Iron.	Sulphuric acid.....	Nitric acid.........	Platinum...	1.177
	Zinc.	Sulphuric acid 1, water 4...........	Nitric acid, fuming.	Platinum...	1.812
	Zinc.	Sulphuric acid 1, water 4...........	" (sp. w. 1.33)	Platinum...	1.678
Grove...	Zinc.	Sulphuric acid 1, water 12...........	" (sp. w. 1.33)	Platinum...	1.603
	Zinc.	Sulphuric acid 1, water 4...........	" (sp. w. 1.19)	Platinum...	1.538
	Zinc.	Sulphuric acid 1, water 12...........	" (sp. w. 1.19)	Platinum...	1.512
	Zinc.	Sulphate of zinc...	" (sp. w. 1.33)	Platinum...	1.550
	Zinc.	Sea-salt, NaCl......	" (sp. w. 1.33)	Platinum...	1.765
	Zinc.	Sulphuric acid 1, water 4...........	Sulphate of copper, concentrated	Copper.....	1.000
Daniell.	Zinc.	Sulphuric acid 1, water 12...........		Copper.....	0.906
	Zinc.	Sea-salt 1, water 4 .		Copper.....	......
	Zinc.	Bichromate of potash 3.	Sulphuric acid 4, water 18....	Copper.....	1 615
	Zinc.			Carbon	1.574
	Zinc.			Platinum...	0.977

IX. ELECTRO-MOTIVE FORCES.

Joule, 1844.—Extract from Wiedemann.

Amalgamated zinc...	Acidulated water...	Sulphate of copper...	Copper... 1.00
Amalgamated zinc...	Solution of potash...	Sulphate of copper...	Copper... 1.88
Amalgamated zinc...	Salted water...	Sulphate of copper...	Copper... 1.06
Amalgamated zinc...	Sulphate of soda...	Sulphate of copper...	Copper... 1.04
Amalgamated zinc...	Acidulated water...	Nitric acid...	Platinum... 1.87
Amalgamated zinc...	Salted water...	Nitric acid...	Platinum... 1.98
Amalgamated zinc...	Solution of potash...	Nitric acid...	Platinum... 2.34

Buff, 1857.—Extract from Wiedemann.

Amalgamated zinc...	Dilute sulphuric acid...	Nitric acid...	Platinum... 2.787
Amalgamated zinc...	Dilute sulphuric acid...	Nitric acid...	Carbon... 1.780
Amalgamated zinc...	Dilute sulphuric acid...	Nitric acid...	Biox. of manganese 1.900
Amalgamated zinc...	Dilute sulphuric acid...	Nitric acid...	Cast-iron... 1.775
Amalgamated zinc...	Dilute sulphuric acid...	Chromic mixture...	Carbon... 1.972

Experiments of Buff, 1865.

Grove...	Amalgamated zinc...	Dilute sulphuric acid...	Nitric acid...	Platinum... 1.750
Bunsen...	Amalgamated zinc...	Dilute sulphuric acid...	Nitric acid...	Carbon... 1.734
Callan...	Amalgamated zinc...	Dilute sulphuric acid...	Nitric acid...	Cast-iron... 1.700
Poggendorff...	Amalgamated zinc...	Dilute sulphuric acid...	Chromic mixture...	Carbon... 1.706
Daniell...	Amalgamated zinc...	Dilute sulphuric acid...	Sulphate of copper...	Copper... 1.000
Meidinger...	Zinc, not amalgama'd	Dilute sulphuric acid...	Sulphate of copper...	Copper... 0.848

Experiments of Beetz, 1853.—Extract from Wiedemann.

Daniell...	Zinc...	Dilute sulphuric acid...	Sulphate of copper...	Copper... 1.000
Grove...	Zinc...	Dilute sulphuric acid...	Nitric acid...	Platinum... 1.708
	Zinc...	Dilute sulphuric acid...	Hydrochloric acid...	Platinum... 1.872
	Zinc...	Dilute sulphuric acid...	Chloride of potassium...	Platinum... 1.505
	Zinc...	Dilute sulphuric acid...	Chloride of sodium...	Platinum... 1.499
	Zinc...	Dilute sulphuric acid...	Bromide of sodium...	Platinum... 1.451
	Zinc...	Dilute sulphuric acid...	Iodide of sodium...	Platinum... 1.021
	Zinc...	Dilute sulphuric acid...	Dilute sulphuric acid...	Platinum... 1.539
	Zinc...	Sulphate of zinc...	Dilute sulphuric acid...	Platinum... 1.466

IX. ELECTRO-MOTIVE FORCES—(CONTINUED.)

Experiments of Petruschefsky, 1857.—Extract from Wiedemann.

Daniell	Amalgamated zinc	SO_3HO, 4 vol. and water 100 v.	Sulphate of copper	Copper	1.030
	Unamalgamated zinc	SO_3HO, 4 vol. and water 100 v.	Sulphate of copper	Copper	0.930
	Amalgamated zinc	Salted water	Sulphate of copper	Copper	1.050
	Unamalgamated zinc	Salted water	Sulphate of copper	Copper	1.010
Grove	Amalgamated zinc	Dilute sulphuric acid	Nitric acid	Platinum	1.780
Bunsen	Amalgamated zinc	Dilute sulphuric acid	Nitric acid	Carbon	1.090
	Amalgamated zinc	Dilute sulphuric acid	Nitric acid	Cast-iron	1.720

Experiments of Raoult, 1864.—"Annales de Chimie et Physique."

Smee	Amalgamated zinc	Dilute sulphuric acid	Platinized platinum		0.59
	When hydrogen is given off upon the platinum				0.69
	Before any freeing of hydrogen, the liquid containing air in dissolution				

Extract from Clark and Sabine's Formula.

Daniell	Amalgamated zinc	Sulphuric acid 1, water 4	Sulphate of copper, satur'd	Copper	1.072
	Amalgamated zinc	Sulphuric acid 1, water 12	Sulphate of copper, satur'd	Copper	0.978
Marié Davy	Amalgamated zinc	Sulphuric acid 1, water 12	Nitrate of copper	Copper	1.000
	Amalgamated zinc	Sulphuric acid 1, water 12	Sulphate of protoxide-of-mercury paste	Carbon	1.524
Leclanché	Amalgamated zinc	Sal ammoniac solution	Sal ammoniac and water	Carbon and MnO_2	1.481
Bunsen	Amalgamated zinc	Sulphuric acid 1, water 12	Nitric acid, fuming	Carbon	1.964
Poggendorff	Amalgamated zinc	Sulphuric acid 1, water 12	Nitric acid (sp. w. = 1.38)	Carbon	1.888
	Amalgamated zinc	Water 100, bichromate of potash 12, sulphuric acid, 25	Sulphuric acid 1, water 12	Carbon	2.028
Grove	Amalgamated zinc	Sulphuric acid 1, water 4	Nitric acid, fuming	Platinum	1.956

X. ELECTRO-MOTIVE FORCES.—General Recapitulation.

Daniell	Amalgamated zinc	Sulphuric acid 1, water 4	Sulphate of copp'r, satur'd	Copper	1.079
	Amalgamated zinc	Sulphuric acid 1, water 12	Sulphate of copp'r, satur'd	Copper	.958
	Amalgamated zinc	Sulphuric acid 1, water 12	Nitrate of copper	Copper	1.000
	Unamalgamated zinc	Sulphuric acid 4, water 12	Sulphate of copper	Copper	.909
	Unamalgamated zinc	Sulphate of zinc	Sulphate of copper	Copper	.955
	Amalgamated zinc	Chloride of sodium 1, water 4	Sulphate of copper	Copper	1.060
Grove	Amalgamated zinc	Sulphuric acid 1, water 4	Nitric acid, fuming	Platinum	1.956
	Amalgamated zinc	Salted water	Nitric acid (sp. w. 1.33)	Platinum	1.804
	Amalgamated zinc	Sulphuric acid 1, water 12	Nitric acid (sp. w. 1.33)	Platinum	1.810
	Amalgamated zinc	Sulphate of zinc	Nitric acid (sp. w. 1.33)	Platinum	1.652
Bunsen	Amalgamated zinc	Sulphuric acid 1, water 10 (?)	Nitric acid	Platinum	1.780
Callan	Amalgamated sinc	Sulphuric acid 1, water 10	Nitric acid	Carbon	1.60
Grove	Amalgamated zinc	Sulphuric acid 1, water 10	Nitric acid	Cast-iron	1.72
Bunsen	Amalgamated zinc	Sulphuric acid	Nitric acid	Platinum	1.759
Callan	Amalgamated zinc	Sulphuric acid	Nitric acid	Carbon	1.734
Poggendorff	Amalgamated zinc	Sulphuric acid	Nitric acid	Cast-iron	1.700
Grove	Amalgamated zinc	Sulphuric acid	Chromic mixture	Carbon	1.796
	Amalgamated zinc	Sulphuric acid 1, water 4	Nitric acid, fuming	Platinum	1.956
Bunsen	Amalgamated zinc	Sulphuric acid 1, water 12	Nitric acid, fuming	Carbon	1.964
Poggendorff	Amalgamated zinc	Sulphuric acid 1, water 12	Nitric acid (sp. w. 1.38)	Carbon	1.888
	Amalgamated zinc	Sulphuric acid 1, water 12	Chromic mixture	Carbon	2.028
Marié Davy	Amalgamated zinc	Dilute sulphuric acid	Sulphate of protoxide-of-mercury paste	Carbon	1.524
Lat. Clark	Amalgamated zinc	Sulphate of zinc	Sulphate of protoxide-of-mercury paste	Carbon	1.33
Leclanché (old mud)	Amalgamated zinc	Saturated solution of sal ammoniac	Sulphate of suboxide-of-mercury paste	Mercury	1.457
De la Rue	Zinc	Saturated solution of sal ammoniac	Bioxide of manganese	Mercury	1.481
Becquerel	Zinc	Sulphate of zinc	Chloride of silver	Silver	1.03
Duchemin	Zinc	Sulphate of zinc	Sulphate of lead	Lead	0.55
Second'y Batt'ry	Platinum	Water acidulated with sulphuric acid	Perchloride of iron	Carbon	1.541
Planté	Lead	Water acidulated with sulphuric acid		Platinum	2.38 1¾ Bun. = 2.53 or 2.94

REMARKS UPON THE PRECEDING TABLES.

It is seen that the Daniell battery may be superior or inferior to the volt, according to the proportions of the mixture of sulphuric acid and water in which the zinc is immersed.

The conclusion to be drawn is that, for a certain composition of this mixture, the electro-motive force of the Daniell is equal to the volt. Consequently, except for researches requiring great precision, the English unit (volt) and the Daniell may be indifferently used.

The figures given by different observers disagree, as is seen from the tables; these differences may arise from various causes, of which the principal is, no doubt, the difference of composition of the mixture in which the zinc is immersed. The table shows that the electro-motive forces of Daniell's and Grove's battery vary with the proportions of the mixture of sulphuric acid and water. It seems as if there might be made a very interesting study upon this question, searching, for instance, the conditions of the maximum electro-motive force of an element.

We conclude our tables with secondary batteries having electrodes of platinum and of lead (Planté). The question of finding out the maximum force of polarization of a voltameter has been studied by several physicists, and it does not seem to have been definitely decided upon. The figures which we give are those of Wheatstone, who was the first to undertake to determine it.

CONCLUSION.

In the preceding chapters we have studied hydro-electric batteries; that is, apparatus capable of producing electric currents by means of chemical energy. They are therefore contrivances which transform chemical action into electricity.

Thermo-electric batteries, of which we have not spoken, are contrivances which transform heat into electricity. They have made very important progress of late years, and everything leads to the belief that before long they will be of great service to industry, whereas up to this time they have only been of interest to physicists.

These two kinds of apparatus are not the only ones by which electricity may be produced; there are to be cited frictional electrical machines, electrophori, and machines like those of Holtz.

But another category, that of magneto-electric machines, has lately assumed a rapidly increasing importance. Enormous progress has been made, notably by Gramme, which has led to the construction of machines of unprecedented power.

These machines, frictional machines, machines of Holtz, magneto-electric machines, are all contrivances which transform movement into electricity, and should, therefore, be classed together.

It appears certain that all the apparatus producers of electricity that will hereafter be invented will be comprised in one of these three classes; in other words, it does not seem possible to produce electricity without expending chemical energy, heat, or movement, because energy only presents itself under these four forms.

Hydro-electric batteries have of late years made less progress than the apparatus of the other two categories, but they are still open to improvement, and will eventually make important progress.

First of all, the exact nature of the chemical reactions which take place in these batteries must be elucidated; it is disgraceful that at the present time it is not known exactly what goes on in Grove's or Bunsen's battery.

The extremely small but incontestable variations in the electro-motive force of even completely depolarized batteries, such as that of Daniell, must be explained.

The variations in the internal resistance must be studied, variations which are but incompletely explained by the change in the chemical composition of the liquids.

Finally, new reactions among the infinite number presented by chemistry must be used, and above all zinc must be dispensed with, which has hitherto thrust itself, so to speak, upon inventors.

www.ingramcontent.com/pod-product-compliance
Lightning Source LLC
Chambersburg PA
CBHW031931230426
43672CB00010B/1886